Basic
Air
Conditioning

Volume 1

Basic Air Conditioning

Volume 1

GERALD SCHWEITZER

A. EBELING
Executive Editor, Environmental Design

HAYDEN BOOK COMPANY, INC.
Rochelle Park, New Jersey

Copyright © 1971

HAYDEN BOOK COMPANY, INC. All rights reserved. This book or any parts thereof may not be reproduced in any form or in any language without permission of the publisher. Library of Congress Catalog Card Number 74-155420

Printed in the United States of America

5	6	7	8	9	10	11	12	13	PRINTING
79	80	81	82	83	84	85	86	87	YEAR

Preface

The components and controls explained in Volume 1 of this book are typical of those used in most modern residential and light commercial air conditioning systems.

Each manufacturer has his own method of applying the various components and controls. No attempt has been made to give detailed information about specific makes and models of equipment, except where such information can be used to enhance the reader's understanding of the material under discussion.

Both volumes together are intended to give the reader a foundation not only in the basic fundamentals of air conditioning theory, but also in the construction and operation of the air conditioning system. In Volume 2 of this series, the fundamentals of application, installation, and service are discussed. Together these two volumes should give the reader a better understanding of the fundamentals, operation, and application of the various air conditioning systems presently used for residential and light commercial applications. In addition, the installation and troubleshooting information provided can be of considerable help to trained technicians.

The authors wish to express their thanks to the many individuals and companies who contributed their talents and efforts to make this book possible. We especially thank Mr. Murray Rosenthal for his technical editing of the manuscript.

Contents

1. Basic Physical Theory 1-3
   What is Air Conditioning, 1-3
   Matter, 1-3
   Temperature, 1-4
   Heat, 1-6
   Pressure and Vacuum, 1-11
   Electricity, 1-16

2. Principles of Environmental Comfort 1-29

3. Principles of Refrigeration 1-32
   Refrigerating Cycle, 1-32
   Basic Mechanical Refrigeration System, 1-33

4. Basic Refrigeration Components 1-36
   Basic Systems, 1-36
   Refrigerants, 1-37
   Evaporators, 1-40
   Compressors, 1-45
   Condensers, 1-61
   Receivers, 1-68
   Refrigerant Metering Devices, 1-69

5. Peripheral Devices 1-83
   Pressure Valves, 1-83
   Pressure Gauges, 1-95
   Filter-Driers, 1-101

6. Motors and Controls 1-105
   Compressor and Fan Motors, 1-105
   Automatic Controls, 1-112

   Glossary, 1-129
   Index, 1-139

Basic Air Conditioning

Volume 1

1
Basic Physical Theory

What is Air Conditioning?

Man has long known the benefits of snow, ice, and cold water in maintaining his food supply. However, it is only since the modern advances in mechanical cooling that man has turned to the problem of enhancing his personal comfort. Hence, the advent of air conditioning.

How does air conditioning differ from refrigeration, its parent? Technically, air conditioning is the control of factors affecting atmospheric conditions surrounding man within a structure; such factors as temperature, humidity, dust, odors, etc. Refrigeration, on the other hand, is basically a process whereby only heat is removed from within a structure. Thus, we see that while they are different, air conditioning and refrigeration have similarities; both are heat-removal processes, with air conditioning being a more sophisticated operation. Since both processes depend on the removal and/or transfer of heat, we should know what heat is, how it is measured, how it is transferred, and what heat may be expected to do under known conditions.

We will therefore need to know the basic principles of matter, temperature and its measurement, and heat and its measurement. To understand the operation of an air conditioner we will also have to know the effects of pressure and vacuum, and how they are measured. Knowledge of the basics of electricity is also necessary to understand how air conditioners are wired internally and to the power source.

Matter

Matter is anything which has weight and occupies space. It exists in any one of three states: solid, liquid, or gas. Regardless of the state of the particular matter, the smallest particle to which it can be broken down and still retain its original properties is the molecule. In each state, these molecules bear a different relationship to each other.

1-4　　　　　　　　　　*Basic Air Conditioning*

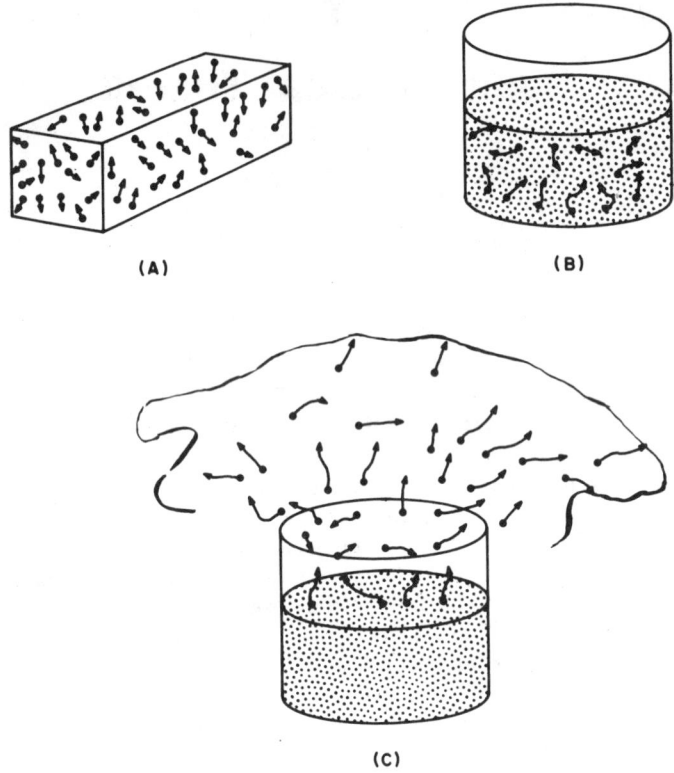

Fig. 1-1. The three forms of matter: (A) solid, (B) liquid, and (C) gas.

In solid form, as shown in Fig. 1-1, the molecules are held closely together so that they are not free to separate from each other. Thus, solids have both definite shape and volume. In liquid form, the molecules are also held together, but are free to move about. Therefore, while a liquid has a definite volume, its shape depends on that of its container. In gaseous form, the individual molecules are not held together. They are free to travel about and become completely dissociated. Thus, gases take the volume and shape of their containers.

Temperature

Definition of Temperature

All substances have two thermal (pertaining to heat) properties: temperature and heat. The temperature of a substance is an indication only of its DEGREE of heat, not the QUANTITY of heat. For instance, a

1-pound piece of lead might have the same temperature as a 100-pound piece, but the QUANTITY of heat in the larger piece would be 100 times greater than that in the smaller piece.

There are three different types of temperature: dry-bulb, wet-bulb, and dew-point.

DRY-BULB temperature is the temperature we are most familiar with—that measured by the common dry-bulb thermometer.

WET-BULB temperature is temperature indicated by a wet-bulb thermometer. Wet-bulb temperature is always lower than dry-bulb temperature.

DEW-POINT temperature is the temperature at which condensation of water vapor begins as the vapor temperature is reduced.

Measurement of Temperature

As just stated, temperature is measured with thermometers. Most thermometers are of the glass stem type, graduated to measure the contraction and expansion of a confined fluid. Two fluids are generally used: mercury (a liquid metal) and colored alcohol. Mercury has been found to be more accurate in the range of $-40°F$ to about $+675°F$, since it expands and contracts uniformly in this range. Lower temperatures can be measured with alcohol-filled thermometers, for which the range is about $-94°F$ to $+248°F$.

The two most commonly used temperature scales are the Fahrenheit (°F) and Centigrade (°C) scales. On the Fahrenheit scale, water boils at 212°F and freezes at 32°F. Thus, the Fahrenheit difference between the freezing and boiling points of water is 180°. On the Centigrade scale, water freezes at 0°C and boils at 100°C. The difference here is 100°, as contrasted to the 180° on the Fahrenheit scale. The relationship between the two scales is expressed mathematically as follows:

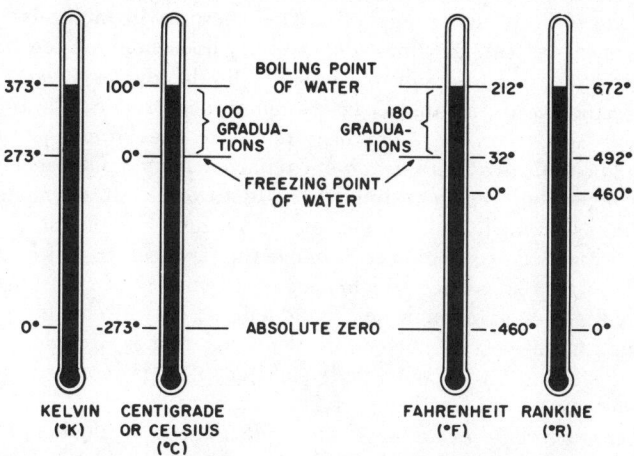

Fig. 1-2. Comparison of temperature scales.

$$C = (F - 32)\,5/9 \quad \text{and} \quad F = 9/5\,C + 32$$

Two other scales used in modern day thermal measurements are the Kelvin (°K) and Rankine (°R) scales. The Kelvin scale is one of absolute temperature, and is based upon the Centigrade scale. Zero degrees Kelvin is equal to $-273.15\,°C$ (usually rounded off to $-273\,°C$). The Rankine scale is also of the absolute type, but is based on the Fahrenheit scale. Zero degrees Rankine is equal to $-459.67\,°F$ (usually rounded off to $-460\,°F$). Figure 1-2 illustrates these relationships.

Heat

Definition of Heat

Heat is thermal energy generated by the movement of molecules in matter. At any temperature above absolute zero ($-273\,°C, -460\,°F$) the molecules of a substance are in motion, and the greater this motion, the greater the heat that is generated. Molecular motion ceases only when the temperature of a substance is lowered to absolute zero, and at this point, the substance is considered to have no heat.

Quantity of Heat

The quantity of heat in a substance is related to molecular velocity; that is, with the rapidity of movement of the molecules. For instance, when electricity is applied to a metal, molecular velocity is increased (due to the electrical force applied to the molecules' atoms, which vibrate increasingly as more electricity is applied). This increase in molecular velocity manifests itself in heat, as witness the ordinary household toaster. Similarly, when heat is applied to a substance (solid, liquid, or gas), the molecular velocity is increased (again, because of increased atomic vibration). The converse is also true: the more heat removed, the lower the molecular velocity. Theoretically, if all heat were removed from a substance, molecular movement would cease entirely. (This does occur, at a temperature of ABSOLUTE ZERO, which is 459.6 degrees below zero Fahrenheit.)

Heat removal or addition can change the physical state of a substance. Adding heat to a substance will first cause expansion of the substance, as shown in Fig. 1-3. If enough heat is added, solids will change to liquids, and liquids to gases. The converse is also true; that is, if heat is removed from a substance, it will contract. If sufficient heat is removed, gases will form liquids, and liquids will solidify.

A characteristic example of these phenomena is alcohol. If enough heat is removed from the alcohol it becomes a solid and has less volume.

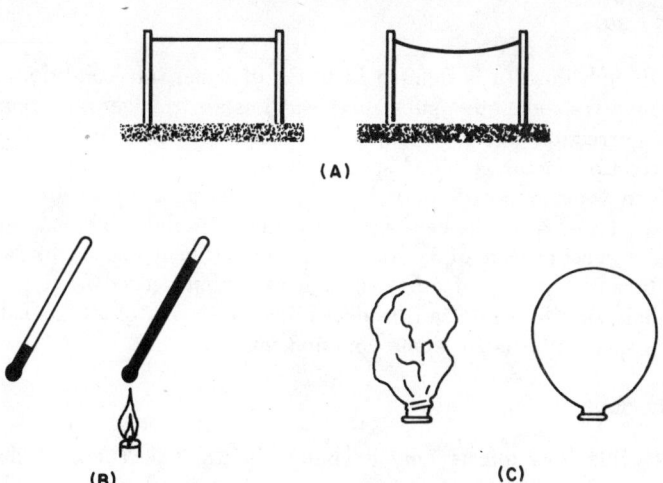

Fig. 1-3. Heat causes expansion of solids, liquids, and gases, e.g. (A) a metal clothesline on a cold and hot day, (B) a thermometer, and (C) a partially filled balloon outside on a cold day and inside a warm house.

Add enough heat and it boils (expands), then becomes a gas. It is this principle of expansion and contraction upon which the thermometer is based.

Measurement of Heat

A definite quantity of heat can be measured, just as can a gallon of water or a bushel of potatoes. By using a substance such as water, which is easily obtainable in a pure form and where characteristics are constant, a standard can be set up as to the amount of heat required to produce a definite change in temperature.

One accepted standard, and the one most used in air conditioning work, is the British Thermal Unit (Btu). One Btu is the heat required to raise the temperature of 1 pound of water 1 degree Fahrenheit.

Another accepted standard is the calorie. One calorie is the heat required to raise the temperature of 1 gram of water 1 degree Centigrade. One calorie is approximately equal to 0.004 Btu. Since the calorie is so small as compared to the Btu, it is seldom used in refrigeration and air conditioning calculations.

Types of Heat

There are three types of heat of interest in air conditioning and refrigeration: specific heat, sensible heat, and latent heat.

Specific Heat

Although the Btu is defined in terms of water, air conditioning frequently involves some other substances, such as air, refrigerants, or metals; hence, a correction factor for the substance(s) involved must be applied. This correction factor is called specific heat.

Since water requires only 1 Btu to raise its temperature 1°F per pound, see Fig. 1-4, it has been used arbitrarily as the standard, and been assigned a specific heat of 1. By comparing the heat needed to raise the temperature of 1 pound of other substances to that needed for 1 pound of water, their specific heats can be determined in terms of Btu's. Table 1-1 gives the specific heats for some common materials.

Sensible Heat

Sensible heat affects only a change in TEMPERATURE; it does not cause a change of state of a substance. It is termed "sensible" because it can be felt by the sense of touch. For example, if 180 Btu's of heat are added to 1 pound of water at 32°F, the result will be 1 pound of water at 212°F, see Fig. 1-5(B).

Latent Heat

Latent heat is that heat added to, or taken from, a substance which will cause a CHANGE OF STATE of the substance, but will not change the temperature of the substance during the time this physical change is taking place. The change of state process may be the change from solid to liquid, liquid to solid, liquid to gas, gas to liquid; it may also be solid to gas or gas to solid directly, without the appearance of the liquid state.

As previously stated, it is the cohesion between molecules that holds solids or liquids together in a compact mass. When sufficient energy to offset the cohesion is supplied (under a given set of conditions), a balance between the cohesive force and the internal energy attempting to disrupt the material is reached at a definite temperature. While the change of state is taking place, all the heat applied is absorbed as energy, which causes the disruption in the molecular structure of the substance. This latent heat,

Table 1-1 Specific Heat of Materials in Btu per lb per °F

Water	1.000
Glass	0.180
Copper	0.093
Alcohol	0.600
Ammonia (liquid)	1.100
Ammonia (gas)	0.520
Lead	0.031
Air	0.240

Specific heats of other materials may be found in standard tables.

Basic Physical Theory 1-9

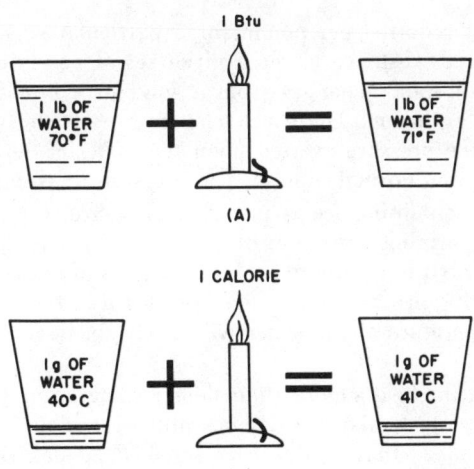

Fig. 1-4. The concept of specific heat: (A) 1 Btu is needed to raise 1 lb of water 1°F and (B) 1 calorie is needed to raise 1 g of water 1°C.

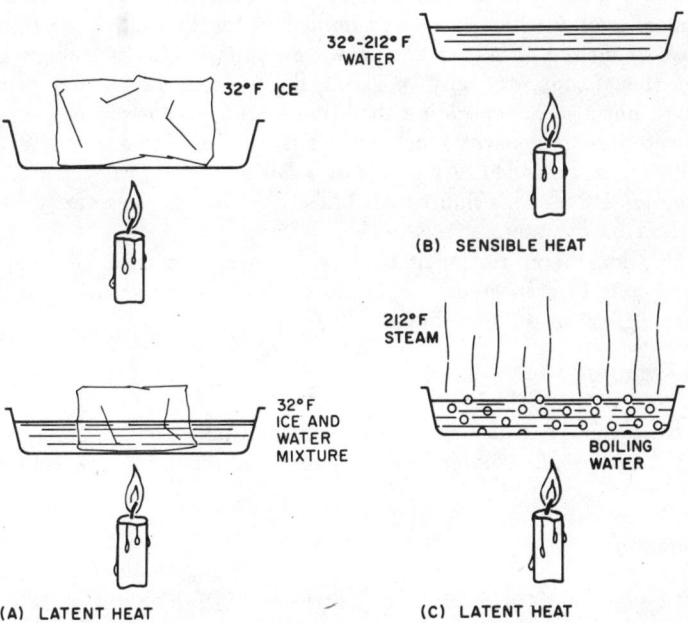

Fig. 1-5. Sensible and latent heat: (A) Heat applied to cake of ice adds latent heat, causing the ice to change in state from solid to liquid. The liquid is at the same temperature as the ice (32°F for a short time). (B) Heat now applied to the 32°F liquid will cause it to absorb sensible heat, raising the temperature to 212°F with no change in state. (C) Continued heating at 212°F adds latent heat to the water, causing a change in state from liquid to gas with no temperature increase.

and the amount required per pound for a particular substance, is always the same for that substance under a given set of conditions.

When a substance changes from a solid to a liquid, the process is called melting, or fusion. The exact temperature at which melting occurs depends upon the pressure exerted upon the material during the process. For this discussion, normal atmospheric pressure is assumed.

If a pan containing ice is placed over a fire, Fig. 1-5(A), the ice begins to melt, forming a mixture of ice and water. This process will continue until the mixture is all water. If the contents of the jar are well mixed during the melting process, a thermometer placed in the water will register 32°F. The temperature of the water will not begin to rise until all the ice has disappeared.

Evidently, the heat supplied has done something quite different from its usual effect of just raising the temperature (sensible heat) of the substance being heated; that is, this heat supplied enough energy to change the molecular structure of the body, thereby changing the state. The amount of heat required to change 1 pound of ice at 32°F to 1 pound of water at 32°F is 144 Btu's. This is the LATENT HEAT OF FUSION of water.

The change of state from liquid to vapor (gas) requires a large amount of energy which must be supplied in the form of heat, just as in the case of melting, Fig. 1-5(C). However, vaporization is a more radical change than fusion, not only in giving the molecules a greater degree of freedom, but also in separating them more widely. Therefore, the heat of vaporization is, in general, much greater than that of fusion. For example, 970 Btu's must be added to 1 pound of water at 212°F to result in 1 pound of steam at 212°F. This figure, 970 Btu's, is known as the LATENT HEAT OF VAPORIZATION of water.

In sum, therefore, latent heat is the heat necessary to change the physical state of a substance; it is not obvious to the sense of touch, and will not register on a thermometer.

Heat Transfer

Heat can be transferred from one substance to another by three different means or combinations thereof: conduction, convection, and radiation.

Conduction

Conduction is a point-by-point process in which the heat transfer is from one molecule to another in a substance, or from one substance to another that is in direct contact.

A typical example of transfer by contact is a finger touching a hot stove. If the temperatures of the two are unequal, heat will flow from the warmer one to the cooler one.

In a like manner, heat travels through the walls of refrigerant tubes and through freezing plates—by conduction.

Convection

Convection is the transfer of heat through motion. Convection involves movement of the substance being heated, and applies to liquids and gases.

A typical example of convection is a stove which heats the surrounding area. Warm air from the stove rises, displacing the heavier cooler air above it. The process continues as long as the stove remains hot; that is, the hot air will continually rise, displacing the cold air, thus transferring the heat through motion.

Radiation

Radiation involves changing heat energy into radiant energy at the source, and the reverting of the radiant energy into heat energy wherever the radiation is absorbed. All bodies radiate heat energy, whether they are hot or cold. The hotter the body, the greater will be its heat radiation.

Radiant heat is energy transmitted by the radiating body. This energy travels in waves. When this radiant energy strikes an object, it is immediately converted into heat again. It is by radiation that the Earth gets heat from the sun.

All bodies receive and reflect radiation, to a greater or lesser degree. Radiant heat will be reflected more readily from a white or polished surface, and will be absorbed more readily by rough or dark surfaces. The reverse is also true; that is, a white or polished surface will radiate less heat than will a dark or roughened surface.

The control of heat and its transfer by mechanical means is built on a foundation laid by the laws of heat just discussed. They must be considered in both the design and application of air conditioning equipment.

Pressure and Vacuum

Definitions

PRESSURE is the force exerted on a substance per unit area, and is usually expressed in POUNDS PER SQUARE INCH (psi). Pressure is calculated by dividing the total force acting on a substance by the total area on which it acts.

For example, if a man weighs 150 pounds, and the area covered by his shoes is 50 square inches, the pressure exerted on this area is 150 pounds divided by 50 square inches, or 3 pounds per square inch (3 psi).

VACUUM is the complete absence of matter; or, to fit our uses, a state of air where the air is so rarefied (thin) that the pressure is far below that of normal atmospheric pressure.

Atmospheric Pressure

Atmospheric pressure is the pressure exerted by the weight of the atmosphere on the Earth's surface. As shown in Fig. 1-6, pressure exerted by the air varies at different elevations. A column of air 1 inch square and extending to a height above the Earth where air no longer exists exerts a pressure of 14.7 pounds on the Earth's surface at sea level. From this can be concluded that the atmosphere surrounding the Earth exerts a pressure of 14.7 psi on the Earth's surface at sea level.

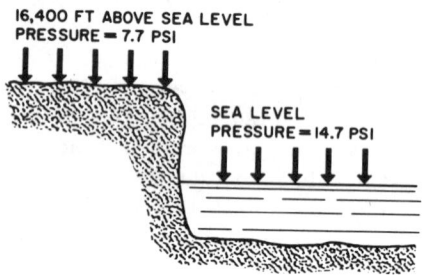

Fig. 1-6. Atmospheric pressure varies with elevation.

If a similar measurement were made at some higher elevation, such as a mountain top, the pressure would be less than 14.7 psi. The height of the air column reaching into the airless space is shorter, and thus exerts less pressure. Atmospheric pressure decreases approximately 1.0 pound for every 2343 feet of increase in elevation.

Absolute Pressure

If all the air were removed from the Earth's surface, the pressure on the surface would be 0, i.e., a complete vacuum. Absolute pressure is measured from a complete vacuum. In a complete vacuum the pressure is 0 pounds absolute pressure. For example, the pressure of the Earth's atmosphere at sea level is 14.7 psi above zero pressure. This pressure is called 14.7 POUNDS PER SQUARE INCH ABSOLUTE (psia). It is also referred to as one atmosphere of pressure, or atmospheric pressure.

Gauge Pressure

In refrigeration and air conditioning work, pressures are usually measured by means of pressure gauges. These gauges are designed to read the pressure above that of the atmosphere; that is, when the gauge reading is zero, the absolute pressure is 14.7 psia, see Fig. 1-7.

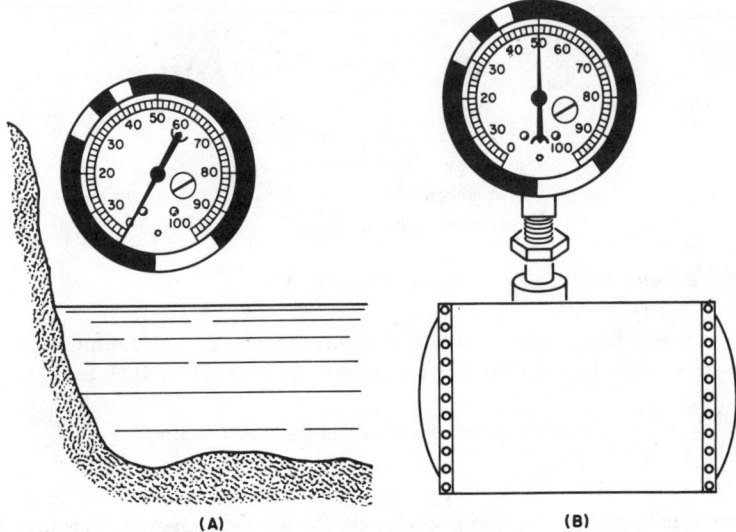

Fig. 1-7. Pressure gauge reads 0 at sea level (A) and reads the difference between actual pressure and sea level (B).

The pressures indicated by a gauge are called gauge pressures, and are referred to in POUNDS PER SQUARE INCH GAUGE (psig). To convert gauge pressure to absolute pressure, add 14.7 to the gauge reading. For example, if a gauge reads 50 psig, the absolute pressure of the system is 50 + 14.7, or 64.7 psia.

Effects of Pressure

The two laws of gases which most affect the fields of air conditioning and refrigeration are Boyle's Law and Charles' Law.

Pressure-Volume Relation—Boyle's Law

Boyle's Law states that the volume of a confined gas at constant temperature decreases in proportion to an increase in absolute pressure and increases in proportion to a decrease in absolute pressure; that is,

$$\text{Volume} \propto 1/\text{absolute pressure}$$

Thus, if a given amount of gas confined in a container is subjected to changes in pressure, its volume will change such that the product of volume multiplied by absolute pressure is always the same.

Mathematically Boyle's Law can be stated as follows:

$$\frac{P_1}{P_2} = \frac{V_2}{V_1} \quad \text{or} \quad P_1V_1 = P_2V_2$$

where

P_1 = original absolute pressure
P_2 = new absolute pressure
V_1 = original volume
V_2 = new volume

Pressure-Temperature Relation—Charles' Law

Charles' Law states that the absolute pressure of a confined gas at constant volume is proportional to absolute temperature; that is,

$$\text{Absolute pressure} \propto \text{absolute temperature}$$

Therefore, if a given volume of gas confined in a container is subjected to a change in temperature, the pressure of the gas will change such that the absolute pressure divided by the absolute temperature is always the same.

Mathematically Charles' Law can be stated:

$$\frac{P_1}{P_2} = \frac{T_1}{T_2} \quad \text{or} \quad P_1T_2 = P_2T_1$$

where

P_1 = original absolute pressure
T_1 = original absolute temperature
P_2 = new absolute pressure
T_2 = new absolute temperature

Pressure and Boiling

Reducing the pressure lowers the boiling point of a liquid; increasing the pressure raises it. For every given pressure acting on a liquid, there is a corresponding temperature at which the liquid will boil.

For example, under conditions of atmospheric pressure (14.7 psia or 0 psig), water will boil at 212°F. If the pressure is raised to 35 psig (49.7 psia), the water will boil at 280°F. If the pressure is lowered below atmospheric until a reading on the gauge of 12.9 in. Hg (mercury) vacuum (6.3 psia) is obtained, the water will boil at approximately 184°F, see Fig. 1-8.

Pressure and Vaporization

Vaporization is the process of changing a liquid to a vapor, either by boiling or evaporation, see Fig. 1-9. While boiling takes place

Basic Physical Theory

1-15

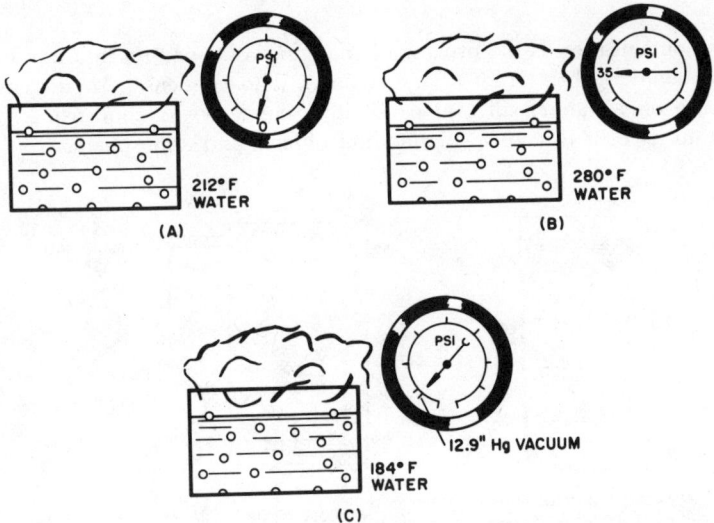

Fig. 1-8. Pressure and boiling: (A) water at 0 psig boils at 212°F, (B) water at 35 psig boils at 280°F, and (C) water at 12.9 in. Hg (vacuum) boils at 184°F.

throughout a liquid, evaporation takes place only at the surface of a liquid. When a liquid evaporates, it absorbs heat from warmer surrounding objects and atmosphere. Reducing the pressure on a liquid lowers the temperature at which evaporation takes place. Therefore, depending on the pressure, more or less Btu's are required to change the state of a liquid.

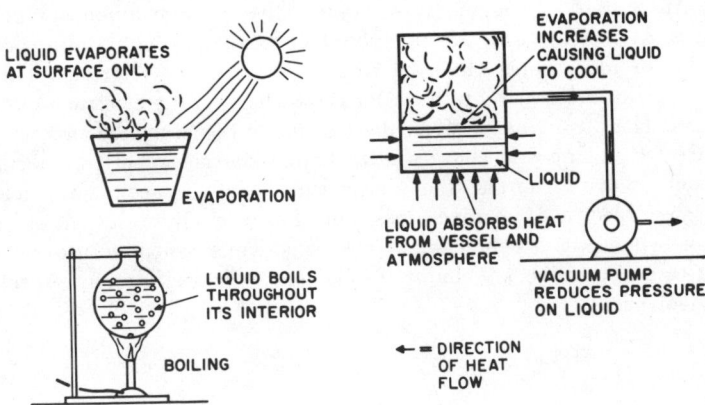

Fig. 1-9. Pressure and evaporation.

Pressure and Condensation

Condensation is the process of changing a vapor to a liquid, see Fig. 1-10. Removing heat from a vapor causes it to condense. An increase in pressure on a vapor will also cause it to condense. A condensing vapor gives up its heat to cooler surrounding objects and atmosphere.

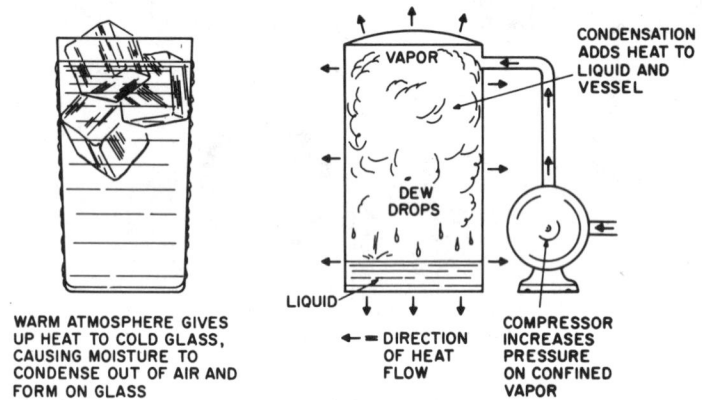

Fig. 1-10. Pressure and condensation.

Electricity

This topic is so vast that proper coverage would require several books. However, only some fundamentals are required to understand the workings of the electrical components in air conditioning and refrigeration systems. Before getting into any detail, two words must be defined in terms of electrical terminology: negative and positive.

All matter is composed of atoms (the basic components of molecules). Atoms, in turn, are composed of a nucleus, containing protons, and one or more rings (like the rings of Saturn) of electrons. The protons are said, arbitrarily, to have POSITIVE charges, the electrons NEGATIVE charges. The attraction between the two holds the atom together (a basic electrical law: opposite charges attract, like charges repel). Returning to our definitions of terms, then, electrical negative means an excess of electrons. Electrical positive does not, however, indicate an excess of protons—it means a deficiency of electrons. Thus, when we use the terms negative and positive, we simply mean more or less electrons, respectively, are involved.

Voltage

Voltage is the force acting on a conducting medium which causes electric current to flow through that medium. Technically, voltage is known

Basic Physical Theory

as potential or electromotive force (emf). However, we are more familiar with the term voltage, and will use it.

Voltage exists in two forms: direct and alternating. Direct voltage is that which causes an electric current to flow in one direction only—from negative to positive. The voltage obtained from a simple flashlight is a typical direct voltage.

Alternating voltage, on the other hand, is that which causes an electric current to flow in different directions, yet still flow from negative to positive. For example, imagine the generator of Fig. 1-11 rotating very slowly (we are looking at the armature head-on). Starting at point 1 (0°), in one time interval it revolves to point 2 (30°); the next interval, it revolves to point 3 (60°); and so on until it returns to point 1 again, having traversed 360°, a complete circle. Now, if we plot these points, as shown in the graph, we have a voltage sine wave; sine because generator action is such that its output is, effectively, dependent upon the sine of the angle between the armature and the north (N) and south (S) magnetic poles within which rotation takes place.

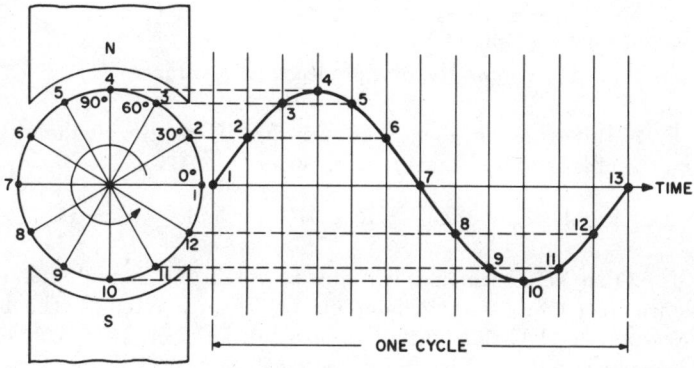

Fig. 1-11. Voltage sine wave.

From the graph it would seem, therefore, that the bottom curve is negative, as it is below the time line (zero voltage line). Not true; what has actually happened is that the voltage has reached a maximum first in one direction (point 4) then reached maximum in the opposite direction (point 10). At the same time, the curve is still going from left to right, i.e., negative to positive, as time goes on. This has been proved by the action of a full-wave rectifier, which converts the alternating voltage to direct voltage, with all the properties of normal direct voltage.

Current

Having defined voltage as the force causing an electric current to flow, what is current? In terms of everyday usage, electric current is the "flow" of electrons in an electric circuit. The word "flow" is in quotes because it is not a flow as we know it, e.g., a flow of water. What actually happens is that the applied voltage excites, or disturbs the electrons nearest it which, in turn, disturb the electrons next in line, and so on. Thus, rather than the electrons flowing, it is the disturbance which "flows."

The currents derived from both voltage sources are direct current (dc) and alternating current (ac). The waveforms for the currents are the same as for the respective voltages. Also, the voltages are commonly abbreviated dc and ac.

Ohm's Law

Before proceeding, it would be well to discuss one of electricity's basic laws, Ohm's Law, which states that the steady electric current (I) in an electrical circuit is equal to the electromotive force (E) divided by the resistance (R) of the circuit. Mathematically,

$$I = E/R$$

where
- I = amperes (A)
- E = volts (V)
- R = ohms–the unit of electrical resistance (Ω)

From this equation, then, it is seen that if any two of the terms are known, the other can be found; that is, since

$$I = E/R \quad \text{then} \quad R = E/I \quad \text{and} \quad E = IR$$

Now that we've demonstrated how to obtain the current, voltage, and resistance in a circuit, what about the power in a circuit? The law for electrical power (P) states that: the power in an electrical circuit is equal to the product of the applied electromotive force (E) and the current (I) flowing in the circuit. Thus, for d-c circuits,

$$P = EI$$

where
- P = watts (W)
- E = volts (V)
- I = amperes (A)

As with Ohm's Law, here, too, if any two terms are known, the other can be found. Hence, if

$$P = EI \quad \text{then} \quad I = P/E \quad \text{and} \quad E = P/I$$

Returning to Ohm's Law, we know that E = IR. So, substituting into the power law,

$$P = EI = (IR)(I) = I^2 R$$

Basic Physical Theory

This latter expression is the true power of an electric circuit, and the term I²R loss, the power dissipated as heat in an electrical circuit, is common.

The preceding discussion was of voltage, current, resistance, and power in d-c circuits. But, in air conditioning, ac is the prime mover. How, then, does Ohm's Law affect a-c circuits?

In terms of basic voltage, current, and resistance, Ohm's Law is the same for ac as it is for dc; that is, $I = E/R$ still holds true. However, power calculations are slightly different. The equation $P = I^2R$ still holds, but $P = EI$ is no longer exactly valid.

To determine $P = EI$ in a d-c circuit, the E is in actual volts, the I in actual amperes. The situation in a-c circuits is different. Rather than the actual volts and amperes, the effective, or root-mean-square (rms), value is used. Refer to Fig. 1-12 and note the waveforms of the two voltages.

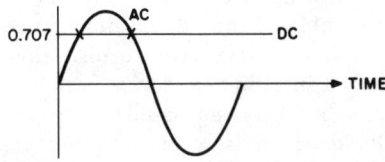

Fig. 1-12. D-c and a-c waveforms superimposed.

The direct voltage waveform is always level, while the alternating voltage is varying. Assuming the same nominal value for both sources, note that the a-c peak rises above the steady d-c level. Now, if it were peak a-c voltage and current used in power calculations, the results would be in error. Peaks are instantaneous, and are not valid for calculations. Thus, for power calculations, the value of the a-c wave that is used is that where the d-c wave intersects—the points labeled X. This is the rms value of the a-c wave, and is equal to 0.707 of the peak level of the wave. For example, if the peak were 120 V, the rms value would be $0.707 \times 120 = 84.8$ V.

Now for the second equation, $P = I^2R$. Again in dc, the actual current and resistance values are used. As for ac, the actual resistance is used, but rms (0.707 of peak) current is used. Thus, when using the $P = EI$ formula for a-c work, one must note the parameters used. If peak values are used for EI, then P is not in watts, but in volt-amperes (VA). If rms values are used for EI, the P is in watts.

Note: The values of voltage, power, and current shown on a nameplate are rms values, and all the following calculations are based on that fact.

Nameplate Voltage

The voltage ratings indicated on nameplates usually have some degree of tolerance. For example, a typical rating in the United States would be 110–120 V ac (rms value). With the standard household voltage being a nominal 115 V ac, the nameplate rating thus indicates that the equipment can operate within a range of ± 5%. Actually, most nameplate ratings permit a range of ± 10% of nominal, from 103.5 to 126.5 V ac. On the other hand, suppose that the nominal source was 220 V ac, a common source for industrial equipment. A ± 10% nameplate rating here means that the associated equipment would operate effectively from 198–242 V ac. However, should the voltage vary above or below these limits, operation would be poor and, in most cases, dangerous; in either case, damage to equipment can occur.

Nameplate Full-Load Amperage-FLA

Current vs Load

The nameplate current rating indicates how much current a device will draw when it is operating under full load. This, in turn, determines what the source must be able to provide. A quick check of the associated fuse and/or circuit-breaker ratings will furnish this information. Said check might also prevent installing a device which draws more current than the fuses or circuit breakers can handle.

On the other hand, there is a limit to the amount of work any device, a motor in this case, can handle safely. If the motor load is such that the current drawn exceeds the FLA rating without blowing the fuse, the result can be disastrous. As explained earlier, the greater the current, the greater the heat generated ($P = I^2R$). Motor windings are composed of varnish- (or other compound) coated wire. Under FLA conditions, these windings normally operate at temperatures 40°C to 50°C above ambient, depending upon the type of motor. However, if these windings draw excessive current, the insulating material can melt from the added heat, causing short circuits and eventual burnout of the whole winding. Thus, the FLA rating is also a motor safety rating.

Then again, in air conditioning and refrigeration equipment the various motors do not always draw exact FLA, because operating conditions may be such as to underload the motors. For example, the belt in a motor-driven pulley might break, leaving the motor running free. Free-running motors can burn out their own bushings and bearings, due to higher friction.

Finally, if a motor is not drawing at or near full-load amperage, its full horsepower is not being utilized. Not only is this a waste of power, but the situation is akin to having an underload—the motor tends to speed up, unless control is provided, and burn itself out.

Current vs Input Voltage

So far, only the effects of the load on FLA have been considered; let us now consider the effects of changes in input voltage.

Basic Physical Theory

Assume that, for some reason, the input voltage drops below the nameplate voltage rating. This has already happened during protracted hot spells in large cities, when air conditioning loads are at their peak. Since the resistance of a motor stays fairly constant (heat increases it slightly), the current drawn by the motor must be less ($I = E/R$). In turn, its horsepower capability is lessened, and the motor cannot operate its load as efficiently as before. Furthermore, many motors have a built-in cooling device, usually a fan on the armature shaft. With the armature turning more slowly, due to the lower voltage-current relationship, the motor can heat up excessively, much like an automobile engine that's been idling too long. The result is armature and/or coil burnout.

Now assume that the input voltage exceeds the nameplate rating. Again, the motor resistance stays constant, and the result is that the motor draws more current, possibly above the FLA rating. Here, too, heat problems arise, due to both excess current and friction. The excess current can cause insulation breakdown and eventual burnout. Friction, due to the increased speed, can cause bushing and bearing burnout. In either case, the motor becomes inoperative. Also with the motor spinning faster, the load itself runs faster, giving rise to additional heat problems and, in the case of a compressor, pressure problems.

Nameplate Frequency

The word FREQUENCY rarely, if ever, appears on a nameplate. Instead, there is usually shown some such indication as CYCLES, with a number next to it or under it; or 50–60 cycles; 60 cycles; 50 cycles, or 60 ∼ (∼ is the graphic representation of the word cycles, and is the shape of a sine wave). Just what is a cycle? In electrical terms, a cycle represents two complete alternations of ac. Thus, the alternating waveform shown in Fig. 1-11 represents one cycle: starting at zero, rising to a maximum, falling to zero again (one alternation), dropping to a maximum in the opposite direction, and then back to zero again (second alternation).

How do cycles relate to frequency? A cycle occurs within a specific amount of time. If just one cycle occurs within this specific time, which is called the PERIOD of the cycle and is, say, one second, then the frequency is one cycle per second. If, say, 60 cycles occur in one second, then the frequency is 60 cycles per second, which is the nominal frequency of almost all voltage sources in the United States. Figure 1-13 shows the relationship between frequency and period of an alternating voltage (or current).

Figure 1-14 compares three different frequences—25, 50, and 60 cycles per second (Hz—hertz—now standard in electronics/electrical work). As shown, the lower the frequency, the fewer the peaks within the same time frame. For example, the figure shows five peaks for 50 Hz and six peaks for 60 Hz, a difference of approximately 16.7%. Nominally, this is not much, and a motor rated at 50–60 Hz will operate efficiently at

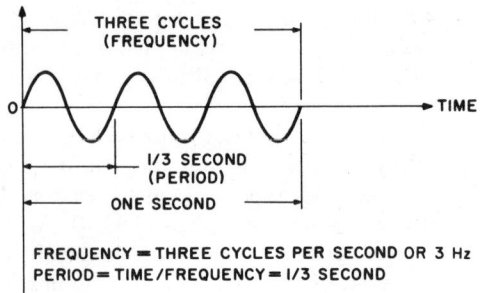

Fig. 1-13. Relation of frequency and period.

either frequency, or in between. However, a motor rated at only 60 Hz will not operate properly from a 50-Hz source—it will revolve at a lower rate (fewer revolutions per minute—rpm), and possibly overheat. The reverse is true for a 50-Hz motor connected to a 60-Hz source; the rpm will be higher, also resulting in overheating and probable burnout. Thus, nameplate ratings must always conform to source frequencies, to ensure safe, proper operation.

Nameplate Phase

On a nameplate, phase is generally listed in second position, e.g., 230 V — 3 PH — 60 Hz. If no phase is indicated, then the device is automatically considered to be a single-phase device.

What is phase? Consider a waveform of a single alternating current Fig. 1-15(A). Assume, for a moment, that the curve represents a roller coaster, with two cars, a and b. Since both cars are connected to the same chain, at any given instant they are both going in the same direction, at the same speed. Hence, they are said to be in phase with each other. Since there is only one roller coaster, i.e., one waveform, this is known as a single-phase system.

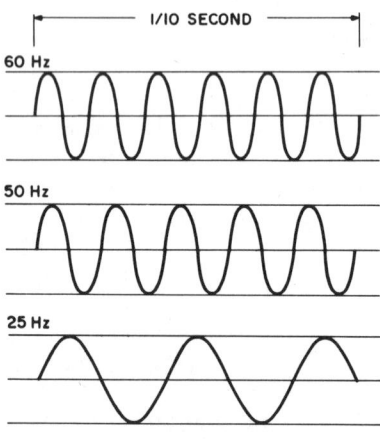

Fig. 1-14. Comparison of frequencies.

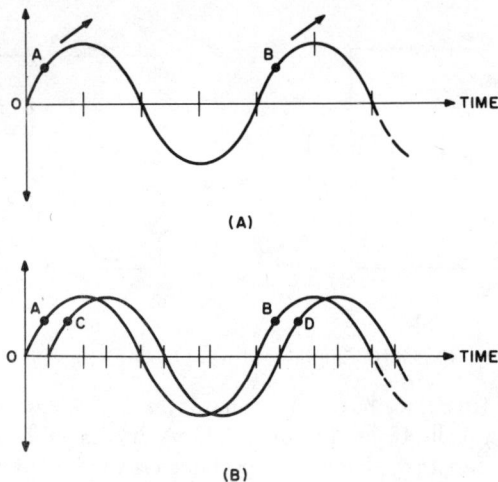

Fig 1-15. Phase representations: (A) single-phase system (B) two-phase system.

Now consider Fig. 1-15(B). Here we have two roller coasters and four cars. As before, cars a and b on the same chain, are in phase with each other. Similarly, cars c and d are on their same chain and are also in phase with each other. But, cars a and c, and b and d are not in phase. Even though they are going in the same general direction, at the same speed, they started at different times. Thus, they are out of phase with each other, and, since there are now two roller coasters, i.e., two waveforms, this is known as a two-phase system.

Why all this fuss about phase? Simply this: air conditioning and refrigeration systems require pumps which, in turn, are driven by electric motors. Fractional horsepower motors can get along on single-phase currents. However, the greater the horsepower, the greater the current drain, and with single-phase current, the greater the drain, the larger the wires needed to conduct it. The result is a large, bulky, expensive motor, plus large, expensive peripheral equipment (switches, circuit breakers, overload protectors, etc.). Also, single-phase motors, without special construction, draw excessive starting currents, so that the wiring might have to be even bulkier.

Two-phase systems, on the other hand, can use smaller windings, smaller wires, etc., because, even though the total current drain is the same, the current is being split into two or more phases, so that each phase is carrying only half or less of the load. For example, consider Fig. 1-16(A), a single-phase system. Here the load (X) is pulling current (I) from the generator through the resistance (R) of one wire and sending it through the resistance (R) of a second wire to ground. The power dissipated is, therefore

$$(R + R)(I^2) = 2I^2R \quad (\text{Ohm's Law})$$

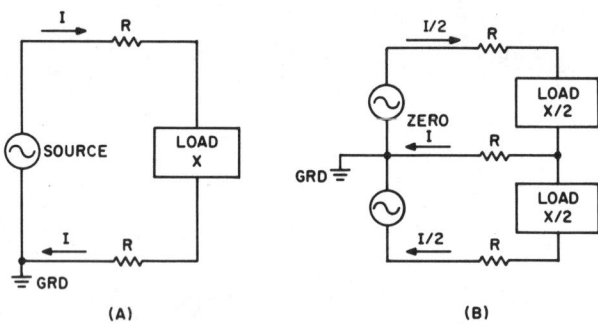

Fig. 1-16. Single- and two-phase circuits.

In Fig. 1-16(B) we have a two-phase, three-wire system. The source, while appearing to be twice that of diagram (A), is exactly the same—it has been split into two phases by special construction of the generator. Now, because the load (X/2) for each phase is only half of what it is in diagram (A), only half the current (I/2) flows through the top and bottom resistances (R). No current flows through the middle R; it is a neutral conductor. The total power dissipated here is, therefore,

$$(R + R)(I/2)^2 = (2R)(I^2/4) = I^2R/2$$

Thus, by utilizing two-phase operation, the power required is cut down to one quarter of what is required for single-phase.

How does this two-phase system translate to an electric motor? Consider, again, Fig. 1-16 (B). Suppose the two loads of the diagram are two separate field windings of a motor. Thus, while one field is excited and pulling the armature toward it, the other is quiescent. Then, as the phase of the source shifts, the second winding is excited and the first becomes quiescent. So, now there are two pulls on the armature, one after the other. Thus, to paraphrase a TV commercial, "not more power, just more power efficiency."

Two-phase systems are generally furnished from 220–230 V a-c sources. Thus, at any one single instant, the voltage in one phase is approaching 220–230 V, while the other is approaching zero.

By adding one more phase and changing the source timing so that the peaks are evenly spaced 120° apart (one complete cycle = 360°), three-phase current results. Three-phase current is used for large loads, and is carried in three conductors. Refer to Fig. 1-17 and note that each line carries one leg of two different phases, so that three such combinations exist. Only one phase can be sensed from any two of the three lines. In other words, if any of the three lines (L_1, L_2, or L_3) is taken away or interrupted, two phases are broken and only a single phase remains.

In summarizing, it can be said that single-phase, two-phase, and three-phase systems are each unique. A single-phase system cannot be used to power a two- or three-phase load. A two-phase system cannot be used to power a three-phase load, or vice versa. However, either a two- or three-phase system can supply single-phase power, providing it meets the voltage and frequency requirements.

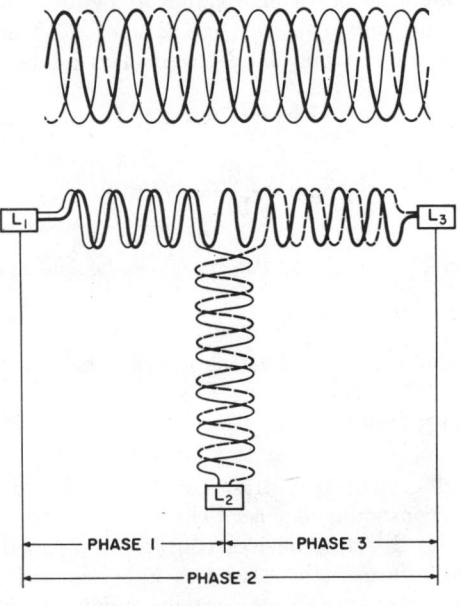

Fig. 1-17. Three-phase circuits.

Power Factor and Efficiency

Power factor (PF) is a figure indicating what portion of the power delivered to a device is actually consumed. In mathematical terms,

$$PF = \frac{\text{power consumed by device (watts)}}{\text{input voltage (volts)} \times \text{current drawn (amperes)}}$$

Thus, to determine the PF of a motor whose nameplate ratings are 115 V ac, 1.5 amperes, 140 watts,

$$PF = \frac{140}{115 \times 1.5} = \frac{140}{172.5} = 0.811$$

In other words, the device is consuming only about 80% of the power delivered.

Efficiency is a measure of just how much work the device can do with the power delivered to it. Efficiency is a percentage measurement and is calculated using the following equation.

Basic Air Conditioning

$$\% \text{ Efficiency} = \frac{\text{rated power output (horsepower)}}{\text{power input}} \times 100$$

Assuming, then, the same nameplate values as before, 115 V ac, 1.5 amperes, and 140 watts, and with, say, a rating of 1/10 horsepower (hp), % efficiency of the motor would be determined as follows:

$$1 \text{ hp} = 76 \text{ watts}$$

Therefore,

$$\frac{1}{10} \text{ hp} = \frac{746}{10} = 74.6 \text{ W}$$

$$\% \text{ Efficiency} = \frac{74.6}{115 \times 1.5} \times 100 = \frac{74.6}{172.5} \times 100 = 0.432 \times 100$$
$$= 43.2\%$$

As demonstrated, the motor is delivering less than half the input to its load.

Power Consuming Devices

All electrical devices that draw current from a power source are considered power consuming devices. Three types exist: resistive, inductive, and capacitive. Resistive devices convert the applied energy to heat (a toaster), or to a form of heat used as light (incandescent bulb). Inductive devices are those which convert the applied energy to heat and some form of mechanical motion (motor), to heat and a change in the applied energy (transformer), or to heat and stored energy (ignition coil in an automobile). Notice that inductive devices always produce heat, due to the current through the coils of wire that comprise the inductive elements.

Capacitive devices do not actually consume power (some large capacitors do get warm, though), they store it; that is, the power builds up in such devices, then is discharged when the applied energy is removed or lowered. Such items are termed condensers or, more commonly, capacitors.

Power consuming devices are usually the loads in electrical circuits. In mechanical circuits, the loads are those devices which usually operate from an electrical device that consumes the applied power, for example, a motor-driven pulley system.

Conductors and Insulators

Conductors are devices used to carry electrical current. Such devices include wire, switches, fuses, circuit-breakers, anything that does not consume significant amount of power. Wire, if it is coiled, can be both conductor and load—it carries current and consumes power in its coiled form (inductive device).

Basic Physical Theory

Insulators are devices that do not carry current in amounts significant enough to affect the circuits they are in. The reason for this is that there are just not enough free electrons in such devices or materials to be excited by the applied power. Some good examples of insulators are wood, porcelain, plastic, rubber, and bakelite.

Some devices, or substances, are known as semiconductors. They will carry current in only one direction, and are the basis for transistors and solid-state rectifiers (diodes).

Electrical Circuits

Electrical circuits comprise conductors, semiconductors, and loads; for our purposes, semiconductors are not involved. Circuits are divided into two basic types: series and parallel.

Series Circuits

Series circuits are those in which all elements are connected directly together across the power source, as shown in Fig. 1-18(A). In such circuits, the CURRENT drawn by each device is the SAME. That is, if the power source is 115 V, and the total resistance (sum of individual resistances of lamp, coil, and heater; fuse resistance is negligible) to the flow of current is 100 ohms, then the total current is

$$I = E/R = 115/100 = 1.15 \text{ amperes}$$

and that same current flows through each of the devices.

Another characteristic of series circuits is that if any of the devices in the circuit opens, e.g., the lamp burns out or the fuse blows, the entire circuit opens and becomes inoperative.

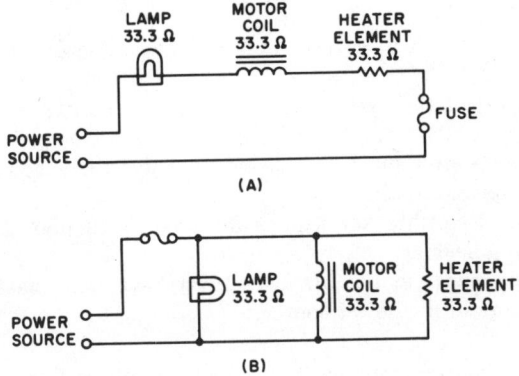

Fig. 1-18. (A) Series circuit and (B) parallel circuit (really series parallel, since fuse is in series with other elements).

Still another characteristic is that the voltage across each device depends upon the device's resistance, since the current through each device is the same and E = IR. The sum of these individual voltages (voltage drops) equals the applied power.

Parallel Circuits

Parallel circuits are those which connect through different paths to the same focal point, in this case, the power source, as shown in Fig. 1-18(B). As opposed to series circuits, the VOLTAGES across each branch of a parallel network are EQUAL to each other and to that of the power source. In other words, if the power source is 115 V, then the voltage impressed on the lamp is 115 V, on the motor coil is 115 V, and on the heater element is 115 V. In a parallel network, it is the current that differs.

Let us assume the same values as before: 115 V and 33.3 ohms resistance for each component. For a parallel circuit, the total resistance is

$$\frac{1}{R_{total}} = \frac{1}{R_{lamp}} + \frac{1}{R_{motor\ coil}} + \frac{1}{R_{heater\ element}}$$
$$= 1/33.3 + 1/33.3 + 1/33.3$$
$$= 0.03 + 0.03 + 0.03$$
$$= 0.09$$

or

$$R_{total} = 1/0.09 = 11.1 \text{ ohms}$$

So, the total current is

$$I = E/R = 115/11.1$$
$$= 10.36 \text{ amperes (almost 10 times that in the series circuit)}$$

The current in each branch is

$$I = E/R = 115/33.3 = 3.453 \text{ amperes}$$

The foregoing calculations tell us that, in parallel circuits:

1. The power source must be capable of delivering much more current than for series circuits.
2. The fuse/circuit-breaker ratings must be much higher, and the circuit wiring heavier.
3. The components in the branches must have nameplate ratings that match those of the power source.

Another important feature of parallel circuits is that if even two of the branches open, the remaining branch will still operate. For example, if both the lamp and heater elements blow, the motor will still operate.

2
Principles of Environmental Comfort

Air conditioning has always been closely associated with mechanical refrigeration and, therefore, many people think of air conditioning as a cooling process. True air conditioning, however, goes beyond mere cooling. In its fullest sense, air CONDITIONING encompasses heating, cooling, ventilating, cleaning, dehumidifying or humidifying, and circulating air for human comfort.

Central air conditioning systems are tied in with heating units—both use the same ducting—and can consist of one large heating-cooling unit, or two separate units.

Ventilating involves the intake of fresh outside air and removal of stale inside air. In the same process, the air is circulated and, by means of filters, cleaned.

As for dehumidifying or humidifying, these are tied in with cooling and heating, respectively. In a good heating system, humidifiers are built in. These units supply a measured amount of moisture to the air. Without this moisture, the warm air might be so dry as to cause discomfort when breathed. Good cooling systems contain built-in dehumidification devices —who ever heard of being comfortable in a cold, damp climate?

Since removal and/or replacement of moisture play such an important role in human comfort, it might be well to discuss humidity and its effects. Humidity can be described as the moisture content of the air. In addition to other gases, air contains evaporated water, called water vapor or moisture. Warm air will hold more moisture than will cold air (colder air tends to condense the vapor). When air contains all the moisture it can possibly hold, it is said to be saturated, and the relative humidity is said to be 100%. If the air contains only half as much water as it can hold at any given temperature, the relative humidity is then said to be 50% If it contains only one fifth of its maximum capacity, the relative humidity is 20%. This amount of water vapor, or relative humidity, affects the perspiration rate.

Basic Air Conditioning

Table 2-1. Conditions for Operation of Air Conditioning Equipment

Type of Room or Building	Outside Dry Bulb Temperature (°F)	Summer Dry Bulb (°F)	Summer RH (%)	Summer Effective Temperature	Winter Dry Bulb (°F)	Winter RH (%)	Winter Effective Temperature	Velocity of Air Over Occupants (FPM)	CFM Per Person	Air Changes Per Hr
Homes	−10 to +10	—	—	—	70	37	65	15 to 25	9 to 13	1.5 to 3
	10 to 30	—	—	—	70	50	66	15 to 25	9 to 13	1.5 to 3
	30 to 50	—	—	—	71	51	67	15 to 25	9 to 13	1.5 to 3
	50 to 70	—	—	—	72	50	67.5	15 to 25	9 to 13	1.5 to 3
	80	78	46	72	—	—	—	15 to 25	9 to 13	1.5 to 3
	85	79	50	73	—	—	—	15 to 25	9 to 13	1.5 to 3
	90	80	51	74	—	—	—	15 to 25	9 to 13	1.5 to 3
	95	*81	52	75	—	—	—	15 to 25	9 to 13	1.5 to 3
	100	*83	50	76	—	—	—	15 to 25	9 to 13	1.5 to 3
Bath	—	80	51	74	80	30	71.5	15 to 25	9 to 13	2 to 4
Kitchen	—	76	52	71	66	50	63	15 to 25	10 to 15	2 to 3

* These summer temperatures are based on intermittent occupancy. Slightly lower temperatures may be desired when occupants remain in the conditioned area. Some people may desire winter inside temperatures slightly higher than those shown.

Normally, perspiration evaporates and, in so doing, absorbs heat, mostly from the body, in the way a refrigerant in a mechanical refrigeration system does when it is vaporized in the evaporator. The ease and rapidity with which this evaporation takes place governs the sensation of coolness. But the ease and rapidity of evaporation is directly affected by the relative humidity of the air. Therefore, from the standpoint of comfort, complete air conditioning should control the relative humidity of the air as well as its temperature.

Controlling the humidity means controlling the vapor pressure, because the difference between the pressure of the vapor exuded by the body and atmospheric vapor pressure determines the amount and ease of evaporation. If the atmospheric vapor pressure is high (high humidity), as it is in the summertime, the body finds it more difficult to evaporate its water. Consequently, less heat is carried away from the body, leaving a feeling of discomfort.

This is why an air conditioning system must be able to reduce the vapor pressure (humidity) in the space to be conditioned until a feeling of comfort is attained. In addition, the room air temperature, which affects convected heat loss, must also be controlled.

Laboratory tests have shown that we can be just as cool at a higher room temperature than otherwise would be comfortable, provided that the relative humidity is reduced. The average person will feel just as cool at a temperature of 79°F when the relative humidity is down to 30%, as he will at 72°F with a relative humidity of 90%. There are, however, practical limits to controlling relative humidity. For human comfort, a relative humidity below 30% would cause unpleasant and unhealthy dryness in the throat and nasal passages. Table 2-1 gives recommended inside relative humidities for both summer and winter comfort.

In summarizing, air conditioning for human comfort does not merely mean making a refrigerator out of a room, so that people can go there to cool off. Air conditioning must not only be able to lower the temperature to a comfortable level, it must at the same time effectively control the relative humidity and maintain adequate circulation.

3
Principles of Refrigeration

Refrigerating Cycle

If we take a liquid refrigerant, confine it in a container, and place this container near a warm object, the liquid will absorb heat from the warm object. If enough heat is absorbed by the liquid refrigerant, it will boil and vaporize. If the vaporized refrigerant gas is sufficiently compressed, it will give up the heat it absorbed from the warm object, and the gas will condense back to a liquid. This process of alternately vaporizing and condensing a refrigerant is called a refrigerating cycle. When the cycle is accomplished continuously through the use of machinery, it it called mechanical refrigeration.

The refrigerating cycle is made possible through the principles discussed in Chapter 1, i.e.,

1. Heat flows from a warmer object to a cooler one. Therefore, objects can be cooled by placing them next to cooler objects.
2. The temperature at which a liquid boils depends upon the pressure exerted on the liquid. The boiling point of the liquid can be controlled if the pressure on it is controlled. Thus, the boiling point of a liquid can be lowered to a point below the temperature of the object to be cooled by the liquid.
3. A boiling liquid continually absorbs heat from any object placed near it. The temperature of the liquid does not rise, as the heat being absorbed is expended in vaporizing the liquid (latent heat of vaporization). Therefore, continuous refrigeration is possible as long as there is liquid available for vaporization.
4. Removing heat from a vapor causes the vapor to condense. Therefore, vapor can be returned to its liquid form after the cooling function has been performed by removing the heat that caused the liquid to vaporize. The temperature at which the vapor will condense depends upon the pressure exerted on it. If the pressure exerted on the vapor is high enough, the temperature of the vapor can be raised to a level where the vapor is hotter than the surrounding air. Heat will then flow from the hot vapor to the cooler surrounding air, and the vapor will condense back to a liquid.

Principles of Refrigeration 1-33

Basic Mechanical Refrigeration System

The basic components required to make up a refrigeration system are:

1. An evaporator—the cooling unit
2. A pump—called the compressor
3. A condenser—the heat disposer
4. A liquid metering device (expansion valve, or capillary tube)

The following describes how these components are put together to form a system.

First, a tube is partially filled with liquid refrigerant. The tube is then placed near the warm object to be cooled. Heat will flow from the warm object to the cool tube. The heat absorbed by the liquid refrigerant in the tube will cause the refrigerant to boil, then vaporize due to its low boiling point. The tube where the boiling takes place is called an EVAPORATOR, Fig. 3-1, since boiling causes one form of evaporation.

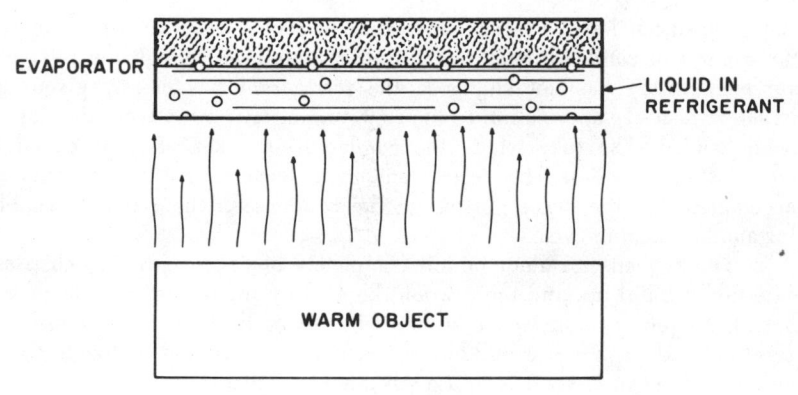

Fig. 3-1. Evaporator absorbs heat.

When enough heat has been removed from the object to be cooled, the next process is to pump out the refrigerant, get rid of the heat the refrigerant has absorbed, and reuse the refrigerant, as it is too expensive to waste. To do this, a pump is added to the system. The same pump is used to pump the vaporized refrigerant out of the evaporator and compress the vapor in another tube. Since the pump is doing the compressing, it is called a COMPRESSOR, Fig. 3-2.

But, the compressor will pump the refrigerant vapor right through the second tube. Therefore, the end of that tube must be blocked. With the system blocked, the compressor can compress the refrigerant vapor, thereby raising the temperature of this vapor high enough, say, approximately 130°F, so that the vapor is hotter than the air surrounding the tube. Heat will now flow from the hot vapor to the cooler surrounding air. When

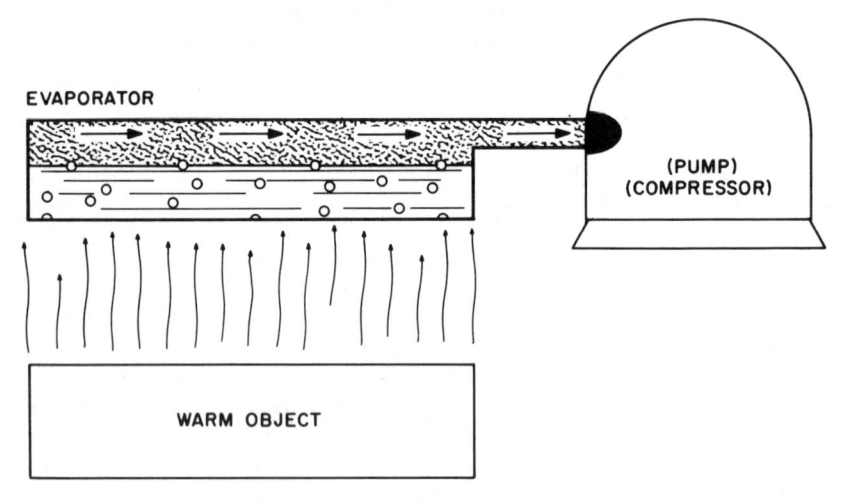

Fig. 3-2. Pump withdraws vapors.

enough heat has left the refrigerant vapor to cool it down to below 130°F, the vapor will condense back to the liquid state, even though the refrigerant temperature has not changed; the refrigerant has merely given up its latent heat of vaporization to the surrounding air, away from the object being cooled. The tube where the condensation takes place is called a CONDENSER, Fig. 3-3. The refrigerant, now back in its liquid state, is accumulated in the lower portion of the condenser, where it is available for another cooling cycle.

The condenser cannot be left completely blocked, as too much pressure might build up, and there would be no way for the refrigerant to get back to the evaporator. Hence, a method must be devised to allow just the right amount of pressure buildup, yet permit recirculation of the refrigerant. The simplest method is to use still another tube.

The laws of physics state that the smaller the diameter of a cylinder, and the greater its length, the greater the resistance it has to the flow of any fluid through it. Thus, by judicious choice of diameter and length, a cylinder or tube can serve as both pressure device and refrigerant carrier. However, such a tube might be excessively long for modern, compact air conditioning units. Hence, it is fashioned into the form of a coil, as shown in Fig. 3-4, thus providing both length and compactness. Such a device is termed a CAPILLARY TUBE. Some systems use various types of valves instead of capillary tubes. These valves are discussed in Chapter 5.

Let us now review the refrigeration cycle, as shown in Fig. 3-5. Compressor action DRAWS vapor from the evaporator, reducing the pressure on the liquid refrigerant in the evaporator. Heat flows from warmer objects to the refrigerant liquid. The reduction of pressure on the liquid causes evaporation, which results in the removal of heat from the liquid, enabling it to absorb more heat from the warmer objects.

Principles of Refrigeration 1-35

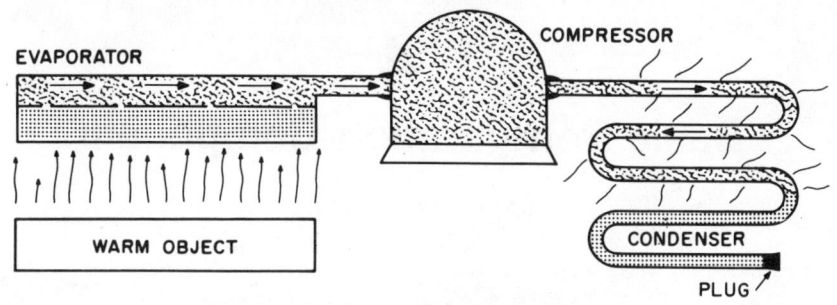

Fig. 3-3. Condenser liquefies vapors.

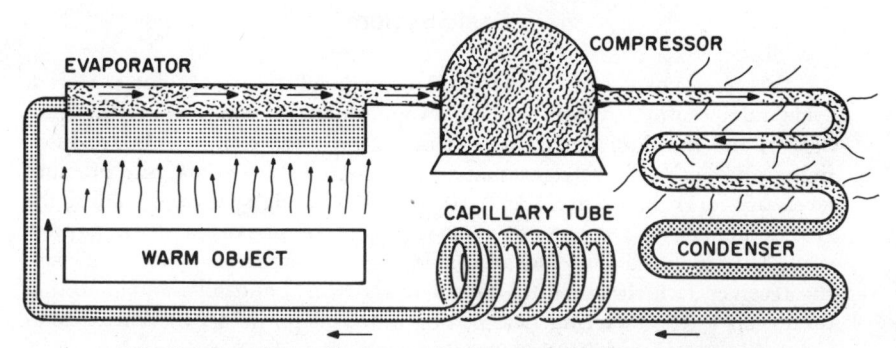

Fig. 3-4. Capillary tube meters refrigerant back to evaporator.

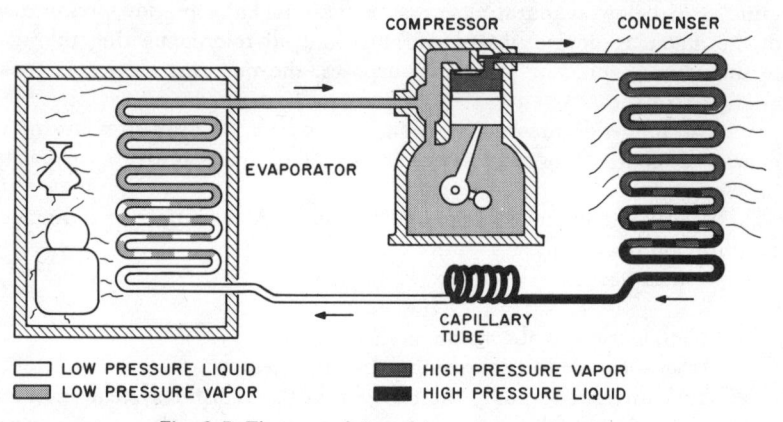

Fig. 3-5. The complete refrigeration system.

The refrigerant VAPOR drawn from the evaporator is compressed to high-pressure vapor, and forced into the condenser. The increase in pressure causes the vapor to condense. The vapor, condensing to a liquid under high pressure, gives up heat to the cooler surrounding atmosphere. The condensed liquid refrigerant is then forced into the capillary tube by the pressure created by the compressor. The liquid from the capillary tube reenters the evaporator, and the cycle continues.

4
Basic Refrigeration Components

Basic System

As mentioned earlier, the basic components necessary for a closed-cycle refrigeration system are the evaporator, the compressor, the condenser, and the liquid metering device. When these components are functioning properly, the refrigeration system operates at two definite pressure levels as shown in Figure 4-1. By definition, the high-pressure side of the system includes all components operating at or above condensing pressure. This includes the discharge side of the compressor, the condenser, the receiver (which stores the liquid refrigerant), and all interconnecting tubing up to the metering device. For practical purposes, the entire compressor is considered HIGH-SIDE equipment.

The low-pressure side of the system includes all components operating at or below evaporating pressure. This includes the low-pressure side of the metering device, the evaporator, and all interconnecting tubing up to the compressor. For practical purposes, the metering device is considered part of the LOW SIDE.

The high-side pressure is commonly known as head pressure or discharge pressure. Low-side pressure is called back pressure or suction pressure.

Briefly, let us once again outline the basic essentials of a refrigeration or cooling system.

It must have:

1. A refrigerant to absorb the undesired heat
2. An evaporator to vaporize the absorbed heat
3. A compressor to pump the refrigerant through the system and simultaneously establish the two pressure levels (suction–low, compression–high) necessary for system operation
4. A condenser to get rid of the heat absorbed previously and re-liquefy the refrigerant
5. A liquid metering device to regulate the refrigerant flow

Another component often used (but not essential in some systems) is a separate receiver used to store refrigerant. Let us examine all of these components and see how they function.

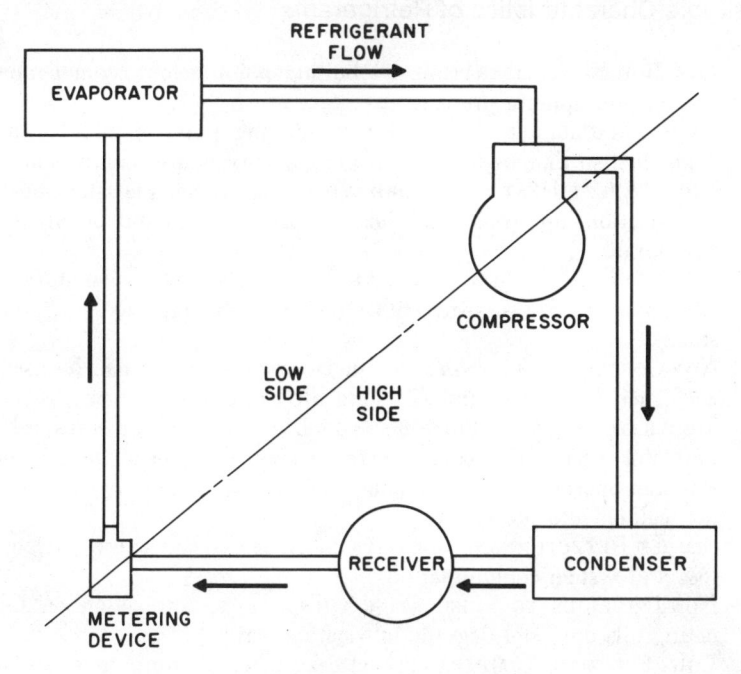

Fig. 4-1. Refrigeration system pressure levels.

Refrigerants

Any fluid that is used as a cooling medium can be called a refrigerant. However, for the purposes of this course (which is limited to residential and light commercial applications) a refrigerant is defined as a chemical fluid whose state can be easily changed from liquid to gas and back again and, when used in a compression system, will cool by absorbing heat during its expansion or vaporization. This type of refrigerant is called a primary refrigerant. Water is not a primary refrigerant; it is merely a carrier of sensible heat.

Primary refrigerants are used in a cooling system to remove heat. The choice of refrigerant used is based upon many considerations. One is the relative ease with which it can be cycled from vapor to liquid and back again. Another is the suitability of its temperature range to the temperature application.

An ideal refrigerant does not yet exist. Even if one could be found, it would never accomplish all the purposes men would use it for. Therefore, any refrigerant used in a system would be a compromise, but, for any particular use, it should have as many desired characteristics as practicable, as described in detail on the following pages.

Desirable Characteristics of Refrigerants

LOW-BOILING TEMPERATURE—a boiling point below room temperature at atmospheric pressure
EASILY MANAGED IN LIQUID FORM—boiling point should be easily controlled so that its heat-absorbing capacity is also controllable
HIGH LATENT HEAT OF VAPORIZATION—the higher the latent heat of vaporization, the greater the heat absorbed per pound of circulated refrigerant
NON-FLAMMABLE, NON-EXPLOSIVE, NON-TOXIC—self explanatory
CHEMICALLY STABLE—to withstand years of repeatedly changing state
NON-CORROSIVE—to assure (1) that common metals may be used to construct the system, and (2) long life to all parts
MODERATE WORKING PRESSURES—high condensing pressures (above 350–400 psi) will require extra heavy equipment construction. Vacuum operation (below 0 psig) introduces the possibility of damp air leaking into the system
SIMPLE DETECTION AND LOCATION OF LEAKS—leaks cause refrigerant loss and system contamination
NON-INJURIOUS TO LUBRICATING OILS—refrigerant action on lubricating oils must not degrade lubricating action
LOW FREEZING TEMPERATURE—freezing temperature must be well below any temperature at which evaporator might operate
HIGH CRITICAL TEMPERATURE—a vapor will not condense at a temperature higher than its critical temperature, no matter how high the pressure. Most common refrigerants have critical temperatures above 200°F
MODERATE SPECIFIC VOLUME OF VAPOR—to minimize size of compressor required
LOW COST AVAILABILITY—to keep price of the equipment within reason and assure proper service when required

Properties of Refrigerants

For air conditioning, a room temperature of approximately 80°F is desirable for comfort. To obtain this temperature, air entering the room from the air conditioner should be in the vicinity of 60°F. Hence, the boiling point of the refrigerant should be close to 40°F, to absorb heat. There are two excellent refrigerants that fulfill the requirements of present-day equipment: Refrigerant-12 (R-12) and Refrigerant-22 (R-22).

REFRIGERANT-12 is chemically known as dichlorodifluoromethane. It was one of the first of the safe refrigerants, and has been used in a multitude of applications including very-large-tonnage centrifugal air conditioning and refrigeration systems, and in household and commercial air conditioning and refrigeration. It is generally used in reciprocating compressors ranging in size from fractional to 100 horsepower and larger. R-12 is ideal for these applications because it condenses at moderate pressures while permitting a broad range of evaporation temperatures. It has a boiling point of −21.6°F at atmospheric pressure. It is practically odorless, and is harmless to breathe except in extremely concentrated form.

REFRIGERANT-22, chemically known as monochlorodifluoromethane, is used in many of the same applicatitons as Refrigerant-12. It is a good low-temperature refrigerant, especially in household applications utilizing reciprocating or rotary compressors. In recent years it has become the most popular refrigerant for residential air conditioning units, because its low boiling point, −41°F, and high latent heat permit the use of smaller compressors and refrigerant lines, which make it ideal for more compact type units. A properly designed compressor will develop up to 60% more

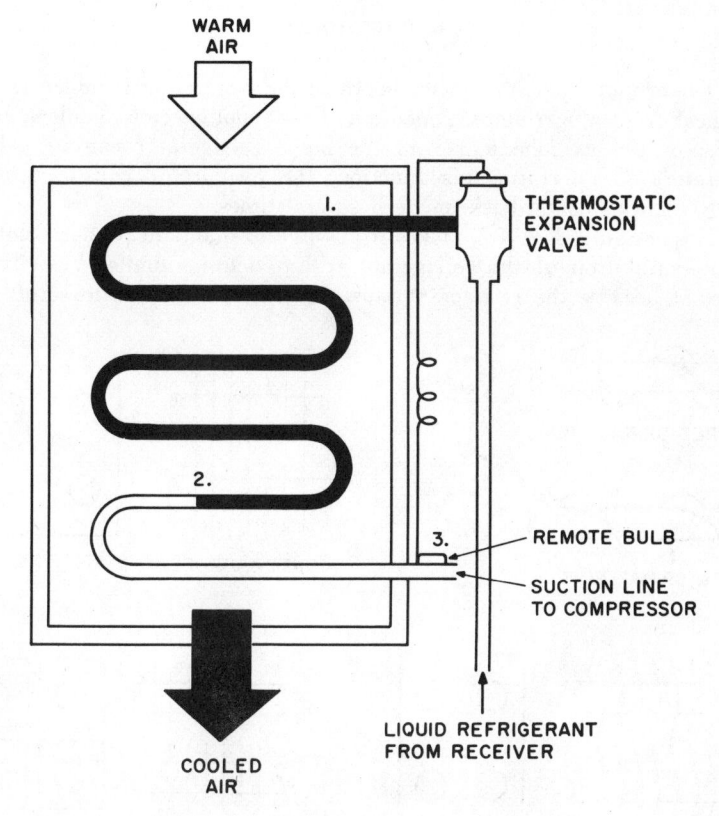

Fig. 4-2. Cooling operation of an evaporator: From 1 to 2, coil is full of liquid and gaseous refrigerant. This part is effective for cooling. From 2 to 3, coil is full of gas only. This part is used to superheat the suction gas.

capacity with R-22 than with R-12. This is extremely important in room air conditioners. Also, R-22 is a good refrigerant for extreme service conditions because it is very stable and has unusually good thermodynamic properties. This is true regardless of whether high capacity or high temperature stability are the major considerations.

Other important properties of R-12 and R-22 are given in the section devoted to charts and tables in Volume 2. The properties of many of the other refrigerants used for various applications are also given in these charts and tables.

In conclusion, let us remember that it is best to use a refrigerant whose critical temperature is well above the design condensing temperature. Too close an approach to critical will result in poor cycle efficiency and very low capacity. A safety margin above the freezing point is needed to avoid the risk of solidifying the working fluid.

Evaporators

The evaporator is the device in which the refrigerant is boiled to extract heat from a surrounding medium. Thus, cooling coils, chillers, unit coolers, or the ice-cube maker in the home refrigerator can be called evaporators. To perform this function, the evaporator coil is usually SERPENTINED, or coiled back on itself several times.

Serpentining, Fig. 4-2, makes it possible to maintain proper velocity and pressure drop of the refrigerant within definite limits. Velocity is important because the refrigerant must have sufficient force to scrub the

CRIMPED SPIRAL FINS

SMOOTH SPIRAL FINS

CONTINUOUS FLAT PLATE FINS

CONFIGURATED PLATE FINS

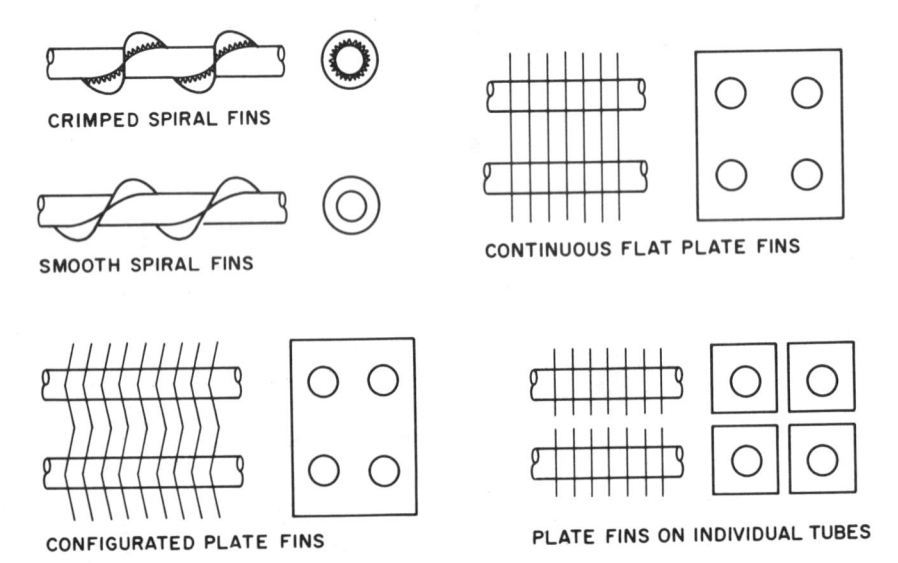

PLATE FINS ON INDIVIDUAL TUBES

Fig. 4-3. Types of fin-coil arrangements.

coil walls, to prevent the film of refrigerant and lube oil from adhering to the walls. The velocity must also be such that the oil moves along continuously with the refrigerant, but not so great as to cause an excess drop in the pressure, since this would create an added load for the compressor.

Fin Arrangements

Evaporators consist basically of various coil or tube arrangements. The tubes may be either bare or have extended or finned surfaces. It is desirable that the coil be effectively and uniformly cooled throughout; hence, finned surfaces are used almost exclusively.

In fin or extended-surface coils the external surface of the tubes is known as the primary surface; the fin surface is the secondary surface. The primary surface consists generally of round tubes or pipes which may be staggered, or placed in line with respect to air flow. The staggered arrangement is usually preferred because it obtains a somewhat higher heat transfer value.

As shown in Fig. 4-3, numerous types of fin arrangements are used, the most common being smooth spiral, crimped spiral, flat plate, and configurated plate. While the spiral fin surrounds each tube individually, the plate type can be either continuous (including several rows of tubes), or individual round or square fins for each tube.

Coil Arrangements

There are basically two types of evaporator coil arrangements, the flooded type and the dry or direct expansion type. The flooded type is designed to carry a constant level of liquid refrigerant within the evaporator, this level being maintained by a float valve or other suitable control. The dry or direct expansion type is designed to handle only that amount of refrigerant actually demanded by the load. Liquid refrigerant is fed to the direct expansion evaporator through an expansion valve in just the right amount so that all liquid will be converted to gas before the refrigerant reaches the suction end of the evaporator.

Flooded Evaporator

The flooded type evaporator, Fig. 4-4, consists of a shell or tank containing a number of short lengths of individual tubes. The tubes are fastened to tube sheets at each end of the shell. Refrigerant liquid is fed through a float valve to the shell and surrounds the tubes. The substance to be cooled flows through the tubes, which are connected together by end sheets in the heads of the shell. The compressor draws the refrigerant vapor from the top of the shell.

Higher rates of heat transfer can be obtained in a flooded type evaporator by using a series of tubes in coil form. Flooded evaporators of

1-42   Basic Air Conditioning

this type use an accumulator or flash tank connected to the tops and bottoms of these coils. The liquid level of refrigerant in the tank is carried high to flood the coils, with the refrigerant feed through a float-operated valve either to the shell or to the bottoms of the coils. The coil tubes are filled with a mixture of water vapor and liquid. Because of the vapor in the tubes, the liquid head (pressure) in the flash tank is greater than the liquid head in the tubes. Therefore, the water circulates by gravity from the bottom of the flash tank, through the tubes and back to the flash tank.

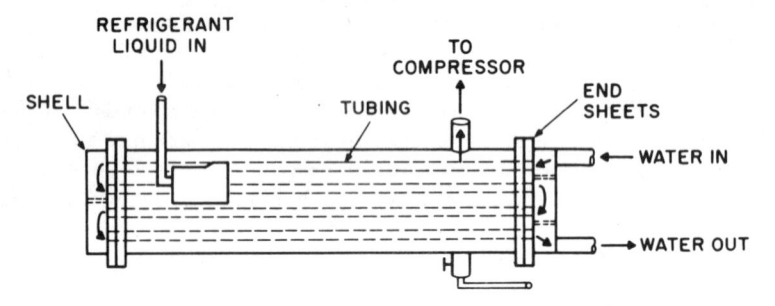

Fig. 4-4. Flooded (shell-and-tube) evaporator.

In a flooded type evaporator considerably more refrigerant is circulated than in a direct expansion type (up to 5 times more). Refrigerant circulation around the tubes depends on the length of the feeds, and the head of liquid imposed upon the liquid inlets. The flow of refrigerant from the liquid inlets must be horizontal and then upward to avoid the possibility of gas trapping.

Flooded type evaporators provide rapid cooling, because the rate of heat transfer from the refrigerated substance to liquid refrigerant is greater than the rate from the substance to the refrigerant vapor.

Dry Evaporators

The dry or direct-expansion evaporator, Fig. 4-5, is the simplest type and is more commonly used in smaller capacity units than are the flooded types. It is known as a dry type because vapor and not liquid occupies a considerable part of the coil volume. It consists of a long length of piping or tubing doubled back on itself to form a coil. The refrigerant from the expansion device feeds into one end of the tube and the suction line connects to the other or outlet end. There is no provision in the direct expansion evaporator to recirculate the refrigerant within the evaporator, or to separate the liquid and the vapor forms of the refrigerant.

To ensure reasonably uniform refrigerant distribution within the evaporator, it is common practice to provide a distributing means, between the expansion device and the coil inlets, in order to divide the refrigerant equally among the feeds. Such a distributor must be effective for both

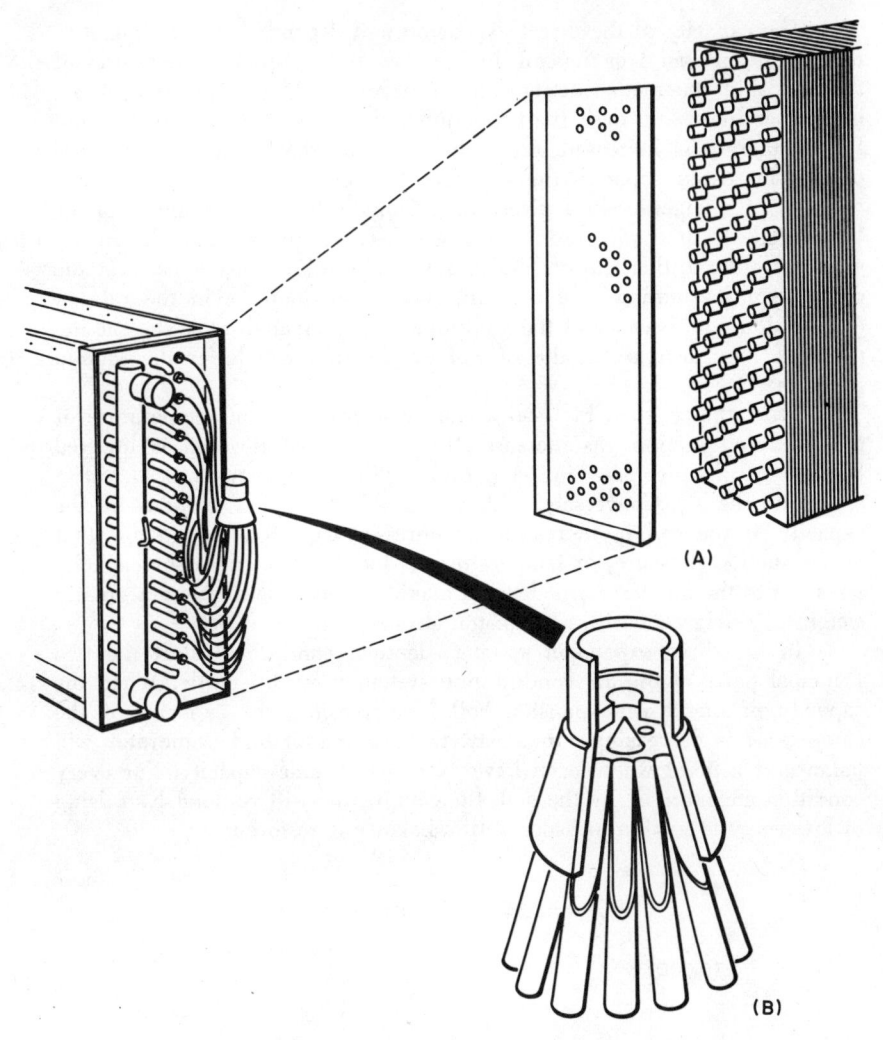

Fig. 4-5. Direct expansion evaporator showing (A) construction and (B) distribution system.

liquid and vapor, because the entering refrigerant is a mixture of the two. These distributor devices will be described more fully later in this chapter.

The inside of the direct expansion evaporator is always filled with a mixture of the refrigerant in liquid and vapor form. At the high-pressure side (inlets), more and more of the liquid is vaporized. At the outlet, the refrigerant is all vapor. The air to be cooled is blown through the evaporator by an evaporator fan. The flow of refrigerant through the evaporator coils is referred to as serpentining.

1-44   Basic Air Conditioning

The capacity of the direct expansion coil depends upon the quantity of air being passed over the coil surface, dry- and wet-bulb temperatures of this air, and the temperature of the refrigerant. If the quantity of air handled is increased, or if the temperature difference between the air and the refrigerant is increased, the coil will have greater capacity. Because of this increased capacity the coil will be able to evaporate a greater quantity of refrigerant in a given time. Conversely, if the quantity of air is reduced, or the temperature difference between the air and the refrigerant is reduced, the rate of refrigerant evaporation will decrease. If the compressor is running at a constant speed, any increase in the rate of evaporation will, because of the additional vapor created, raise the suction pressure. Any decrease in the rate of evaporation will lower the suction pressure.

From the chart in Fig. 4-6 it can be seen that if the temperature of the saturated suction gas increases, the capacity of the coil is lowered because of the decrease in temperature difference between the refrigerant and the air. Conversely, with a decrease in suction temperature the capacity of the coil in increased. Compressor capacity tables show that as the suction pressure, or temperature, is increased, compressor capacity rises due to the increased gas density, making it possible to pump a greater weight of refrigerant with each piston stroke.

In the direct-expansion system, identical conditions affect the two principal parts of the air conditioning system very differently. When the capacity of the direct-expansion coil is increasing, the capacity of the compressor is decreasing. Obviously, the compressor and evaporator will balance at a point where each have exactly the same capacity. For every condition encountered by the coil, the compressor will respond by raising or lowering the suction pressure until a balance is restored.

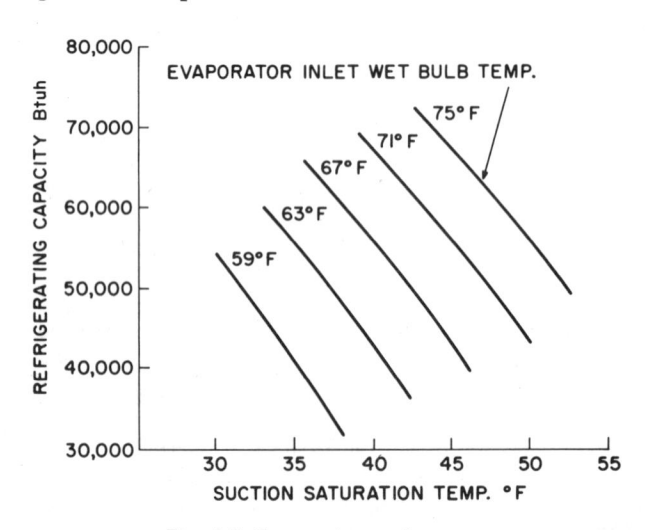

Fig. 4-6. Evaporator performance.

Compressors

The compressor is often called the heart of the refrigeration system, Fig. 4-7. It provides the system with the force to draw the vapor from the evaporator, force it into the condenser, and keep this circulation going. Stated briefly, the low-temperature vapor from the evaporator flows through the suction line to the compressor. The compressor compresses the low-temperature vapor, raising the pressure and temperature. The hot, high-pressure vapor then flows to the condenser where the vapor gives up its heat and is condensed. Because of the resulting lower temperature, heat flows from the space to be cooled into the evaporator, and vaporizes more of the liquid refrigerant. The vapor containing the absorbed heat flows to the compressor, where it is compressed and the temperature raised. Then the high-temperature vapor, containing the absorbed heat from the evaporator, is discharged to the condenser, where the heat flows from the hot vapor into cooler air or water around the condenser.

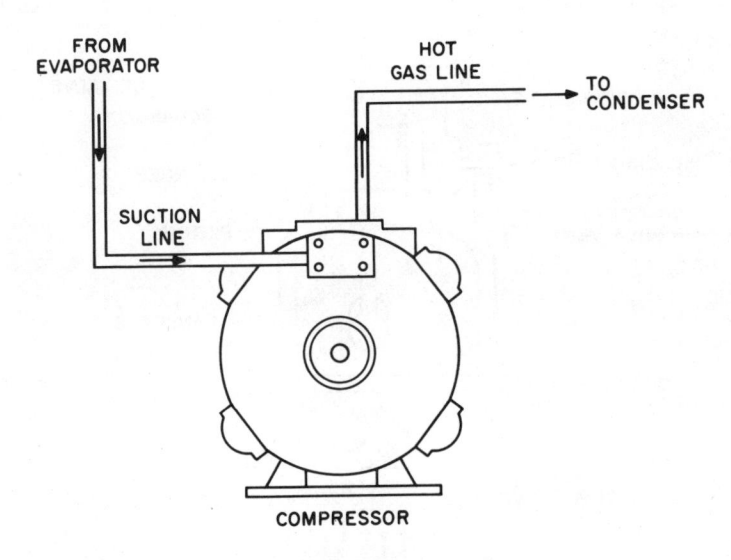

Fig. 4-7. Compressor system.

There are two main groups of compressors: positive-displacement and dynamic. Positive-displacement types are those which compress a fluid within a closed space. Dynamic types are those with rotating vanes or impellers which take in the fluid then expel it at great force. Under the former category fall the reciprocating and rotary-type compressors; under the latter fall the centrifugal compressors.

Reciprocating Compressors

Reciprocating compressors are capable of moving fluids at up to 100,000 cubic feet per minute (CFM) at up to 35,000 psi, depending upon the number of stages employed. These compressors are classified in accordance with cylinder design, compressor drive (direct or indirect), valve design, lubrication, and cooling (air or water). Basically, however, classification depends mostly upon cylinder design.

Single-Acting Compressors

Single-acting compressors are those whose cylinders have pistons which have only one end acting on the fluid, and perform only one function at a time—either suction or compression. Figure 4-8 shows a typical single-acting system. The piston has only one-sided action (only its top is closed), and only one action is performed on each upstroke or downstroke of the piston. On the downstroke, the discharge valve is closed and the suction valve is open, allowing the vapor to enter the cylinder from the suction

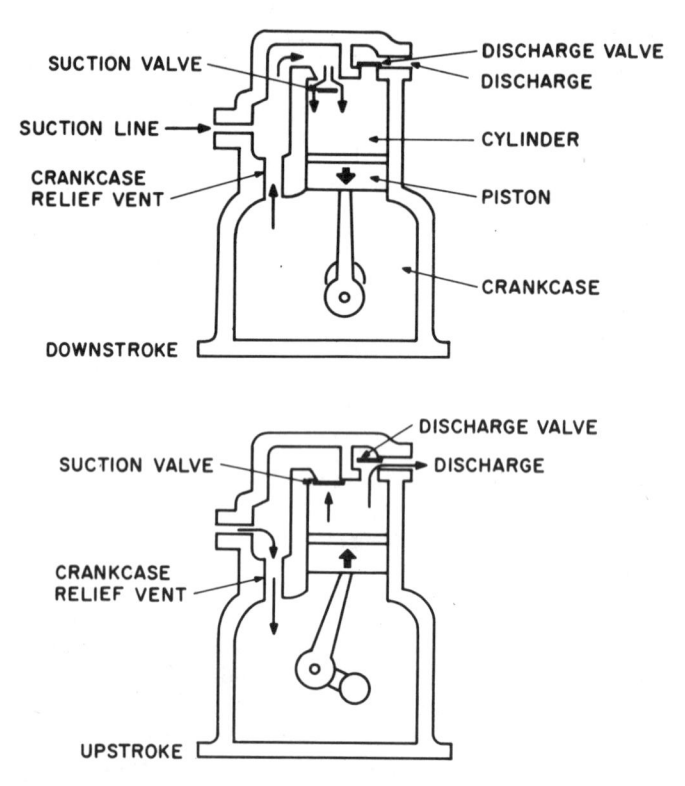

Fig. 4-8. Operation of a single-acting reciprocating compressor.

line and the crankcase relief vent. On the upstroke, compression forces the suction valve closed and the discharge valve to open, forcing the vapor out through it. The crankcase relief vent acts to prevent pressure buildup in the crankcase during the downstroke and the creation of a vacuum during the upstroke. Note that the single-acting system uses only one suction valve and one discharge valve. Single-acting pistons are also called single-trunk pistons.

Double-Acting Compressors

Double-acting compressors are those whose cylinders have pistons with both ends acting on the fluid, and perform two functions simultaneously—suction AND compression. Thus, on the upstroke, a double-acting piston (also known as a double-trunk piston) sucks in vapor from one end of the cylinder while compressing vapor in the other. The reverse occurs on the downstroke. For double-acting cylinders, two suction and two discharge valves are required.

Compression Stages

Compression can be performed in single-, double-, or multistage operation. Single-stage compression is that where only one cylinder operates, with the output from its discharge valve feeding directly to the fluid container, in our case the condenser.

Two-, or double-stage compression is that where the discharge from one cylinder is fed into the suction valve of a second cylinder, where additional compression is accomplished. The discharge from the second cylinder (second stage) is then fed into the fluid container.

Multistage compression is just an expansion of two-stage compression, with three or more cylinders in operation.

Compressor Mounting

Reciprocating compressors are mounted in a variety of configurations. They can be mounted vertically, as in Fig. 4-8. Where two-, or multistage operation is required, each cylinder is mounted vertically, one behind the other, similar to an inline gasoline engine.

Horizontal mounting is that where the cylinder lies on its side. In other words, an upside-down "L" arrangement. For two-stage operation, picture a "T" arrangement, with the two cylinders forming the crosspiece, and the crankcase forming the base.

"V"-type arrangements are also used. Here the cylinders form a "V" on top of the crankcase so, in effect, the whole thing looks like a "Y."

Another arrangement is the semi-radial type. Here, both a "V" and two horizontal cylinders are arranged around the crankcase, in a sort of fan arrangement.

Lastly, there is the angle-type mounting. Here, at least one vertical and one horizontal cylinder form the arrangement. The term "angle" is used because the two cylinders form a right angle, and because the piston rod of the horizontal cylinder is also angled; the rod forms a very obtuse angle instead of being straight.

Open and Closed Compressors

Reciprocating compressors are also subdivided on the basis of drive, and whether they are open or closed. Drives are either indirect, i.e., the crankshaft is turned by an external source such as a pulley-belt arrangement; or direct, with the crankshaft connected directly to a motor.

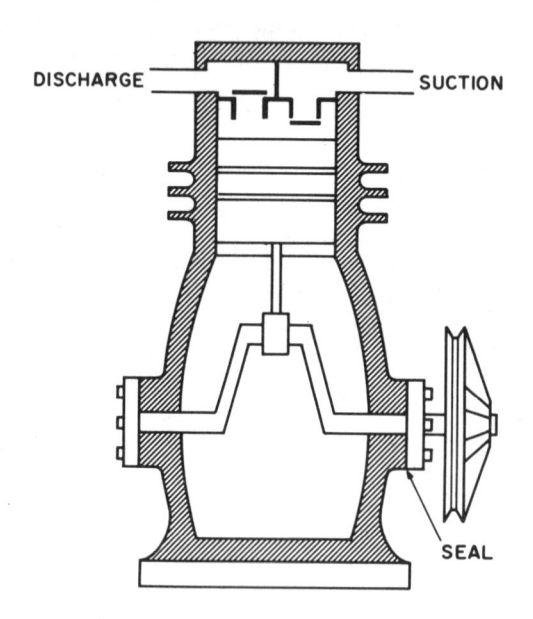

Fig. 4-9. Open-type reciprocating compressor.

Open-type compressors, such as that shown in Fig. 4-9, are indirectly driven. Closed, or hermetic types, are usually direct-driven, with the motor sealed within the compressor housing, as shown in Fig. 4-10.

Closed-type compressors are usually classified as either serviceable or non-serviceable. The former can be taken apart so that individual components can be repaired and/or replaced. The latter are welded shut, so that servicing in the field cannot easily be performed. This disadvantage is offset by the fact that lubrication oil cannot leak out, as it can in the serviceable type.

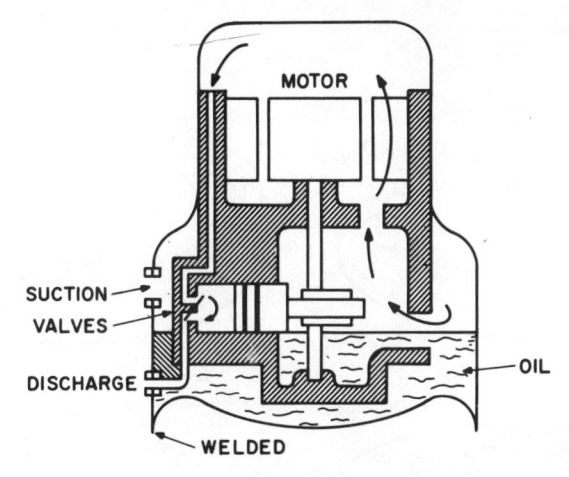

Fig. 4-10. Closed or hermetic reciprocating compressor.

Valves

There are two basic types of compressor suction and discharge valves in common use: the reed or disc type, and the ring type, Fig. 4-11. Although mechanically-operated valves might have an advantage, they have proven unsatisfactory for compressor use, because each change in evaporator or condenser pressure requires a change of valve setting.

The REED VALVE is a thin piece of spring steel, with a portion of the valve covering either the suction or discharge port of the compressor. The spring tension within the steel tends to keep the valve closed. The valve is forced open by pressure which overcomes the spring tension.

RING VALVES are made of heavier metal formed in the shape of rings. These valves are normally held in the closed position by springs. When the pressure under the valve becomes greater than the spring tension, the valve lifts from its seat.

All refrigeration compressor valves depend for their operation upon a difference in pressure between the inside of the compressor cylinder and the suction or discharge pressure. The pressure differential required for valve operation depends upon valve design and compressor speed.

Seals

The end of the open-type compressor through which the crankshaft projects is commonly called the seal end. A seal is necessary at this point to prevent leakage. The four most common types of seals are the packing gland, the stationary bellows, the diaphragm, and the rotary type seal.

The PACKING GLAND seal, Fig. 4-12, generally consists of metal, asbestos, and graphite packing inserted in a recess around the end of the crankshaft. The packing is held in place and compressed around the shaft

1-50   *Basic Air Conditioning*

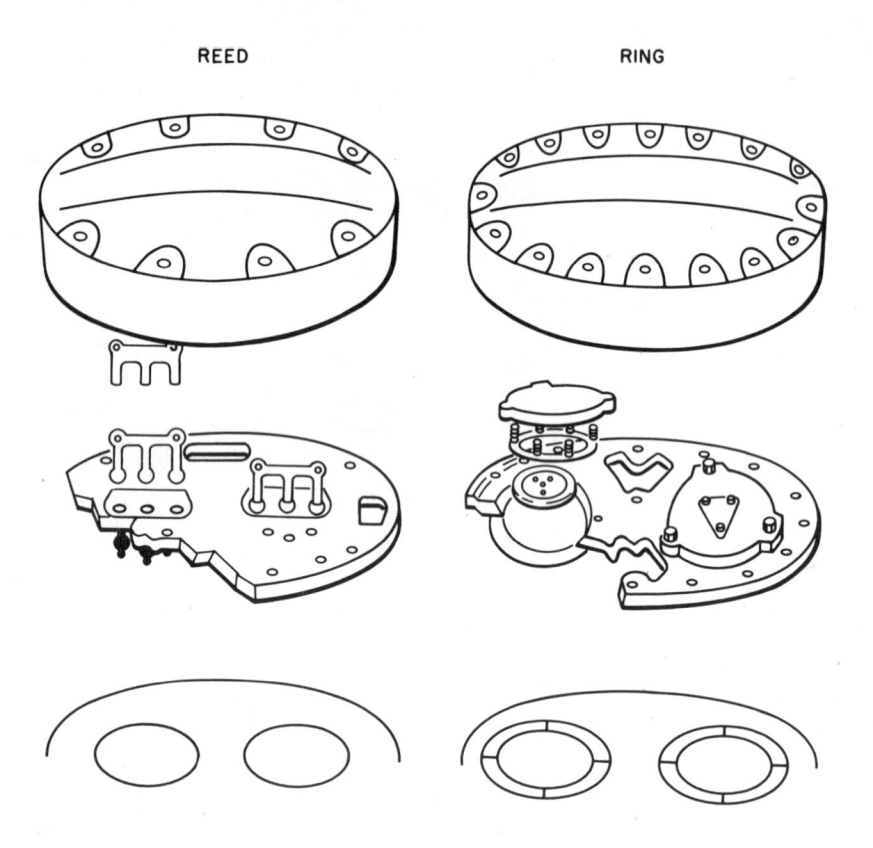

REED   RING

Fig. 4-11. Basic types of valves.

by a metal gland and packing nut. Some installations require a spring between the nut and the gland to compensate for wear. This type of seal is used largely in compressors that operate at relatively low speeds.

The STATIONARY BELLOWS type seal, Fig. 4-13, has a metallic bellows and ring, or nose, backed up by a spring to force it against a shoulder on the crankshaft. The bellows and ring are fixed to the cover plate and do not rotate with the shaft. The sealing surfaces are between the ring on the bellows and the shoulder on the shaft.

The DIAPHRAGM seal, Fig. 4-14, is composed of a stationary diaphragm with an attached nose that presses against a shoulder or metal collar sealed to the crankshaft. A fulcrum is provided to force the nose of the diaphragm against the crankshaft shoulder.

The ROTARY seal rotates with the crankshaft to which it is attached. The nose of the seal rotates against the polished surface of the seal cover plate. A spring keeps the proper pressure between a carbon seal ring and the cover plate.

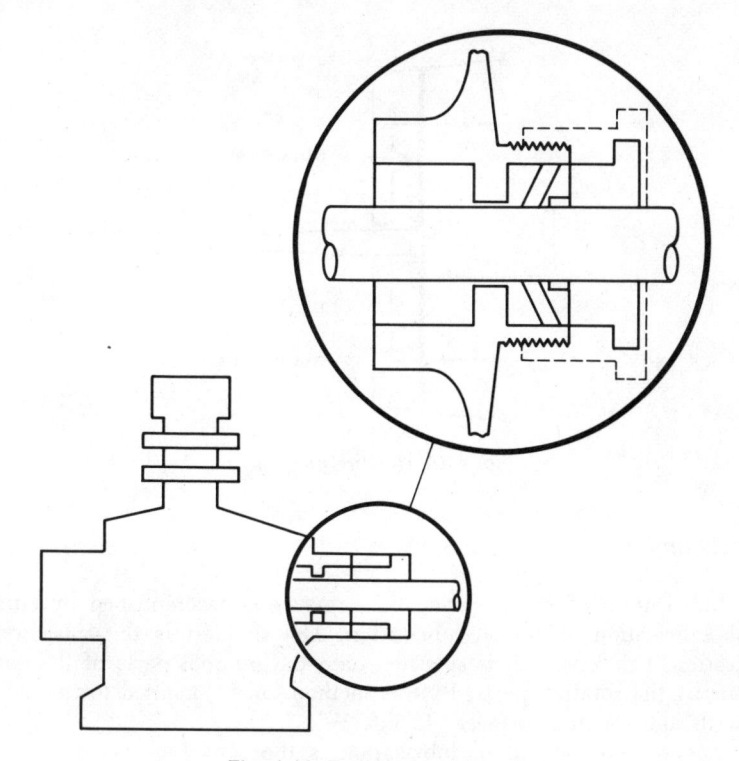

Fig. 4-12. Packing gland seal.

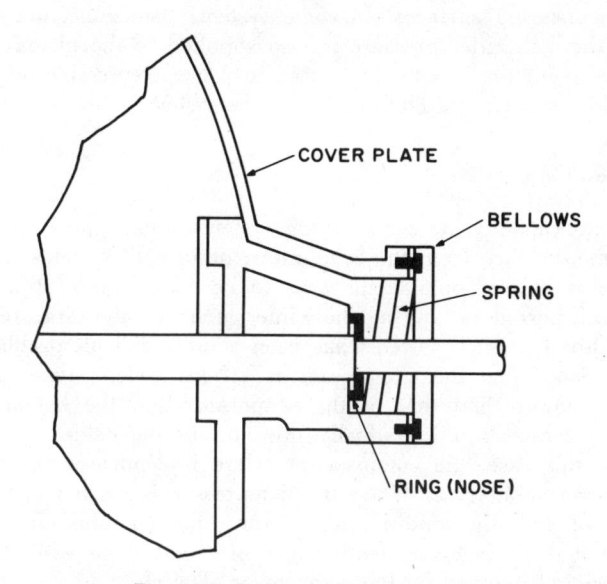

Fig. 4-13. Stationary bellows seal.

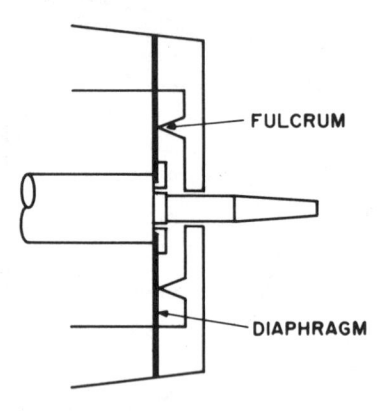

Fig. 4-14. Diaphragm seal.

Lubrication

Lubrication of reciprocating compressors is accomplished by either splash lubrication or forced lubrication. The simplest is the splash-type lubrication. Crankcase oil is splashed onto the internal parts of the compressor by the rotating parts. By this method, oil is supplied to the cylinder walls and bearing surfaces.

The second method of lubrication is the forced or pressure type. This system utilizes an internal pump that forces oil through the crankshaft to the bearing surfaces. In some systems the connecting rods are drilled so that oil under pressure is also supplied to the piston pins. An oil pressure regulator is sometimes used to prevent excessive oil pressure which could result from a malfunction in the system.

Liquid Slugging

In large-capacity air conditioning systems, where the compressor is located some distance from the evaporator, or in split systems, where the compressor is located outside the area to be cooled and subjected to a prolonged off period (as during the winter months), the large refrigerant charge required in such systems has been a most difficult problem to all manufacturers. When the compressor is off for a long time, and at a temperature lower than that of the evaporator and the balance of the system, refrigerant liquid gradually migrates to the coldest part of the system (in this case, the compressor). Here it condenses in the oil of the compressor crankcase. When the compressor is started again at the beginning of the air conditioning season, this combination of liquid refrigerant and oil foams violently, and oil is entrained with the refrigerant vapor and pumped by the compressor. The chart in Fig. 4-15 illus-

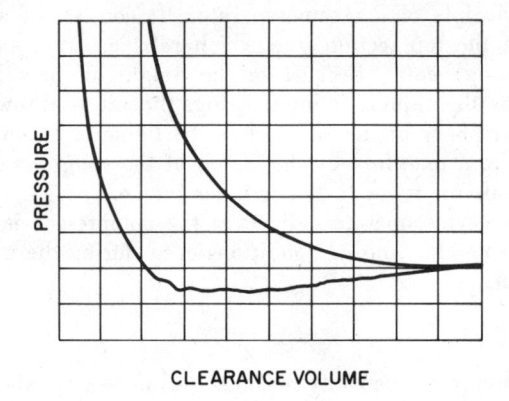

Fig. 4-15. Compressor liquid slugging.

trates the effect of this oil (liquid) SLUGGING in the compressor. Liquids do not compress. As a result, when the clearance volume (volume between discharge valve and top of cylinder) is filled with liquid, the pressure will increase dangerously. The sharp rise of the curve at the end of the stroke indicates the extreme pressure encountered. Slugging usually causes noisy operations and can lead to other undesirable conditions, such as:

1. Loss of compressor capacity due to re-expansion of liquid within the cylinder during the suction stroke
2. Increase of input power required
3. Possible physical damage to the compressor

This condition crops up in any compressor design. The most common way to eliminate it has been to use a heating coil or crankcase heater to keep the compressor warm and prevent the refrigerant from accumulating in the compressor crankcase. The difficulty here is that it is relatively easy for someone to inadvertently disconnect the heater and completely eliminate any protection.

Another possible solution is the PUMP-DOWN CYCLE, where a service man, at the end of each season, pumps all the liquid into the receiver and shuts off valves on either side of the receiver. This however, requires another service call when start-up is required at the beginning of a new summer season.

The most foolproof method is the ANTI-SLUG DEVICE, Fig. 4-16, which operates on the principle of a cream separator. It consists of a rotating ring with holes on the top section into which refrigerant vapor and oil (if the oil is foaming) enter. Because of the weight of the oil and the centrifugal effect of the rapidly spinning ring, the oil is thrown out of the holes on the periphery of the ring, while the lighter refrigerant gases are sucked inward to a manifold by the action of the compressor pistons, and travel down collector tubes to the cylinder intake.

The anti-slug device operates whenever the compressor is running, which means it can handle unusual conditions even during the normal air conditioning season.

Capacity Control

Control of compressor capacity is sometimes necessary where refrigeration or air conditioning loads are variable. In this case, the compressor is called upon to do only a portion of the work for which it was designed; that is, the load is not as great in fall as in summer. When the system operates under partial loads, suction pressures are low. This may result in coil freezeup.

Capacity control may be obtained by: containing suction pressure, controlling discharge pressure, returning the discharge gas to suction, adding re-expansion volume, opening a cylinder discharge port to suction while closing the port to the discharge manifold, reducing compressor speed on open-type units, and closing off the cylinder inlet. The most commonly used are: opening the suction valves by some external force (cylinder unloading), bypassing hot gas discharge to suction within the compressor, and hot gas bypassing external to the compressor.

CYLINDER UNLOADING, Fig. 4-17, is accomplished by holding the suction valve open, thus preventing compression of the gas. A popular method of accomplishing this is by incorporating a hydraulically-operated valve-lifting mechanism. When full capacity is not needed, suction pressure is reduced, and an external solenoid opens, allowing oil pressure to be relieved from a capacity reduction piston. This piston raises the lift pins which lift the suction valve from its seat. The compressor piston is then no longer able to compress the refrigerant, and merely moves up and down within the cylinder. With this method, compressor capacity is limited only by the number of cylinders. Also, the motor horsepower required decreases in almost direct proportion to capacity reduction.

The CYLINDER BYPASS system, Fig. 4-18, is another method of compressor capacity control. This system is activated automatically by either a temperature or a pressure control, but can also be activated manually. When the control calls for capacity reduction, the solenoid opens, and the discharge gas from cylinder Block B passes directly to the suction line. Because the check valve does not allow high pressure gas to enter Block B, and the lines are sized quite large, no high pressure is created in Block B. Consequently, Block B (the bypassed cylinders) operates at suction pres-

Basic Refrigeration Components

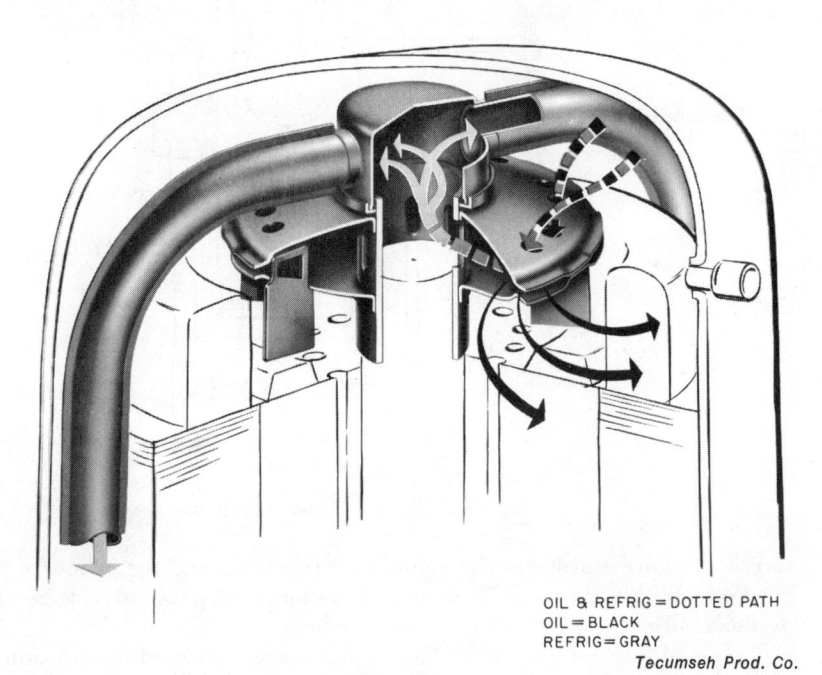

1-55

OIL & REFRIG = DOTTED PATH
OIL = BLACK
REFRIG = GRAY

Tecumseh Prod. Co.

Fig. 4-16. Anti-slug device.

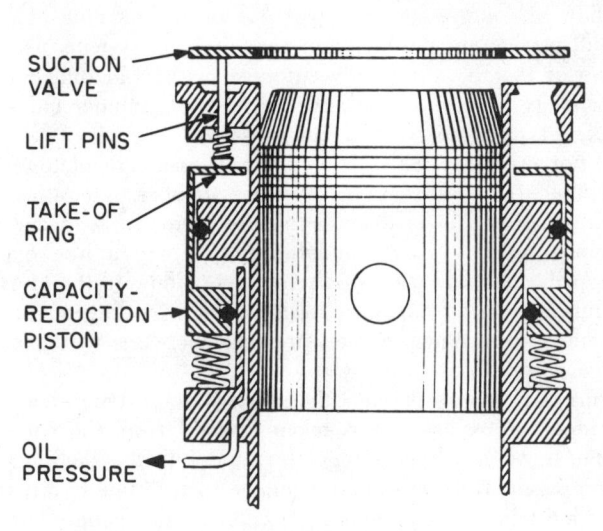

Fig. 4-17. Cylinder unloading.
Trane

1-56   Basic Air Conditioning

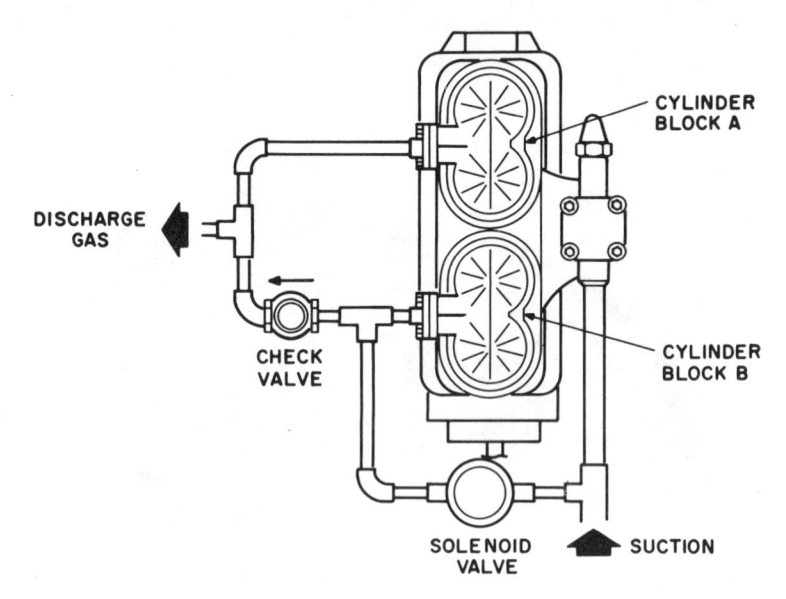

Fig. 4-18. Cylinder bypass system.

sure both above and below the cylinder valve plates, and the cylinders do not work. With this method, the motor horsepower required also decreases in almost direct proportion to capacity reduction.

One of the best and by far the simplest method devised for unloading either large, small, open or hermetic compressors is the HOT-GAS-BYPASS method. This method employs a modulating control valve to provide a metered flow of compressor discharge gas to the low side of a system. It permits full refrigerant flow (output capacity) on systems using any type of compressor that is not already equipped with unloading valves, and it extends capacity reduction below the last step of cylinder unloading range in unloading type multi-cylinder compressors.

The hot-gas-bypass valve is a suction pressure regulator which opens whenever there is a decrease in suction pressure. Therefore, when the suction pressure begins to drop because of reduced load conditions, the hot-gas-bypass valve (also a modulating type) in a line bypassing the condenser will open, allowing discharge gas to flow to the low side of the system, thus increasing suction pressure and providing an artificial load for the compressor. When suction pressure increases to the bypass valve setting, the valve closes.

Figure 4-19 shows perhaps the most common type of hot-gas-bypass system. Here, the bypass line is taken directly from the compressor discharge line, through a bypass regulator, and into the suction line leading to the compressor. Hot gas should enter the suction line in a manner which provides good mixing with suction gas entering the compressor. Prolonged

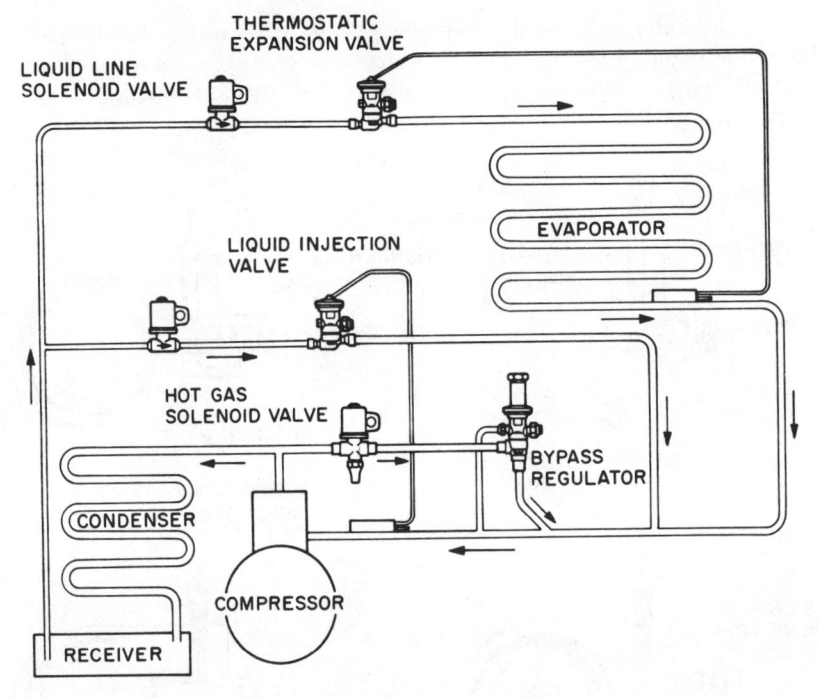

Fig. 4-19. Hot-gas bypass system.
Alco Valve

bypass operation in this system may necessitate the use of a liquid injection expansion valve to control the refrigerant temperature, i.e., keeping it from getting too hot.

Rotary Compressor

Rotary compressors are capable of moving fluid at up to 50,000 CFM, but only at very low pressures, less than 15 psi. Multistage operation is required for higher pressures.

Rotary compressors are widely used for fractional-tonnage applications. They run quietly, have limited vibration, and can be used where a fairly high volume of refrigerant per ton of cooling must be moved. The materials of construction are very similar to those for the reciprocating type, but operation is different; the pumping effect is produced by a rotary rather than a reciprocating motion.

There are two principal types of rotary compressors; one uses sealing blades that rotate with the shaft, and the other has a stationary blade, but the blade has a rubbing contact against the rotating shaft. Operation of the stationary-blade type, Fig. 4-20, is a simple process. The shaft rotates in a cylinder with the eccentric ring on the shaft constantly rubbing against the outer walls of this cylinder. As the shaft turns, the stationary blade imprisons quantities of gas which are gradually compressed into a smaller space, building up the pressure and temperature, and finally forcing the gas into the high pressure side of the system.

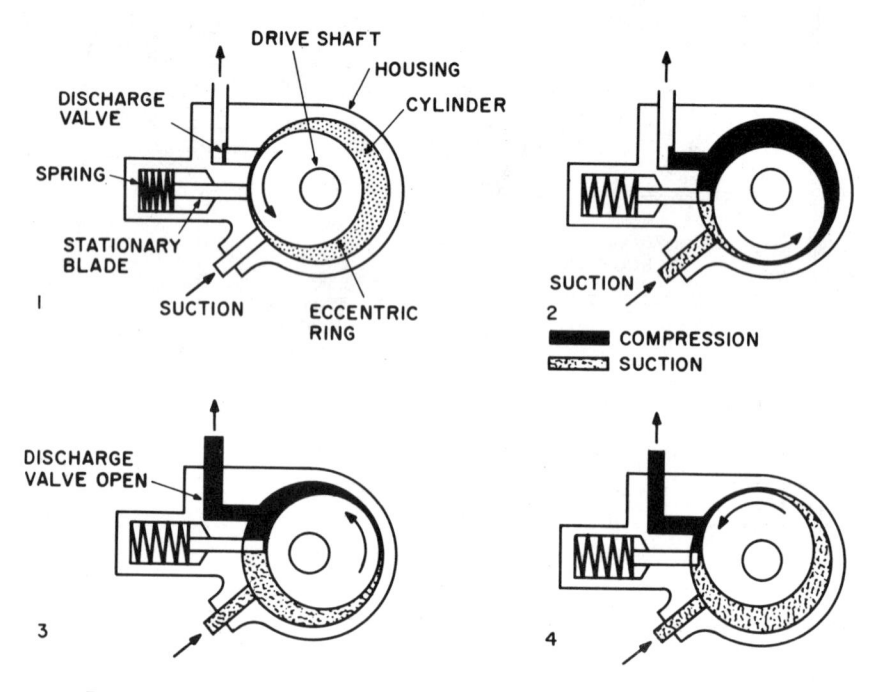

Fig. 4-20. Operating procedure of a stationary blade rotary compressor.

In the rotating-blade type, Fig. 4-21, low-pressure gas enters the cylinder and is imprisoned between two or more blades as the blades rub against the wall of the cylinder. As the rotor revolves, the low-pressure gas gradually compresses into a high-temperature, high-pressure condition. As the gas, now occupying very little space, comes opposite the exhaust opening it passes into a high-pressure dome. It cannot go through the point of contact between the rotor and the housing, nor can it back up because of the following rotating blade. This type of rotary compressor does use an eccentric ring, but the rotor is mounted off-center enough to allow it to almost rub against the housing at a point between the intake and exhaust ports.

Because of the small clearances involved, trouble-free operation of rotary compressors depends upon maintaining a continuous film of oil on the cylinder, rotor, and blade surfaces. This oil feeds into the cylinder through main bearings. Sometimes, a force-feed lubrication system or separate oil pump is used.

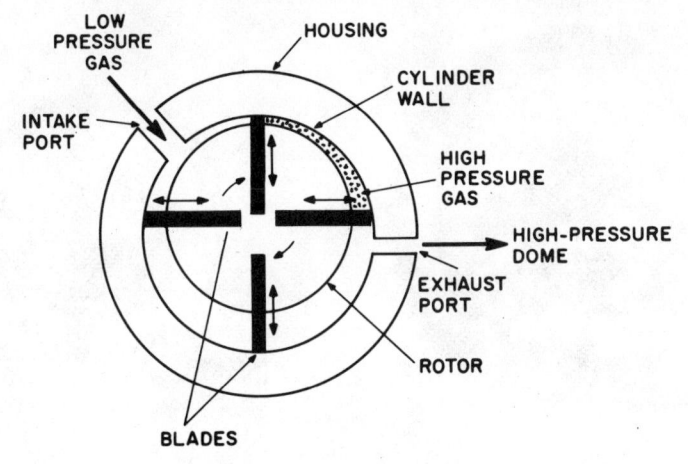

Fig. 4-21. Rotating blade compressor.

Centrifugal Compressors

Centrifugal compressors are capable of moving fluids at up to 150,000 CFM at up to 5000 psi, depending upon the number of stages. They are particularly well-suited for refrigerants with high specific volume. Because of the simplified lubrication, they are frequently used for extremely-low-temperature applications.

They are adaptable to a wide temperature range; from $-130°F$ to $+50°F$. Most of these units are used primarily for large-capacity machines, not for the majority of units discussed here. However, the information is valuable.

Centrifugal compressors compress the fluid by means of centrifugal force. They operate in two different modes: axial flow and rotating impellers. Axial-flow compressors move the fluid in a direction parallel to the rotating axle. They comprise pairs of fixed and rotating blades, with each pair comprising one stage. The rotating blades fling the fluid at high velocity against the fixed blades, thus obtaining compression. The pressure rise per stage is small; hence, many stages are required, and even these go only as high as 150 psi.

The impeller types also use velocity as the compression factor, but have higher compression capabilities. Two types are common: radial-blade and backward-curved impeller.

Radial-Blade Compressor

Radial-blade compressors, Fig. 4-22(A), fling the fluid out at extremely high tip velocities into the passage between the flat walls of the diffuser section. This generally creates a higher pressure for a given blade-tip speed, which allows this compressor to operate at higher speeds than for

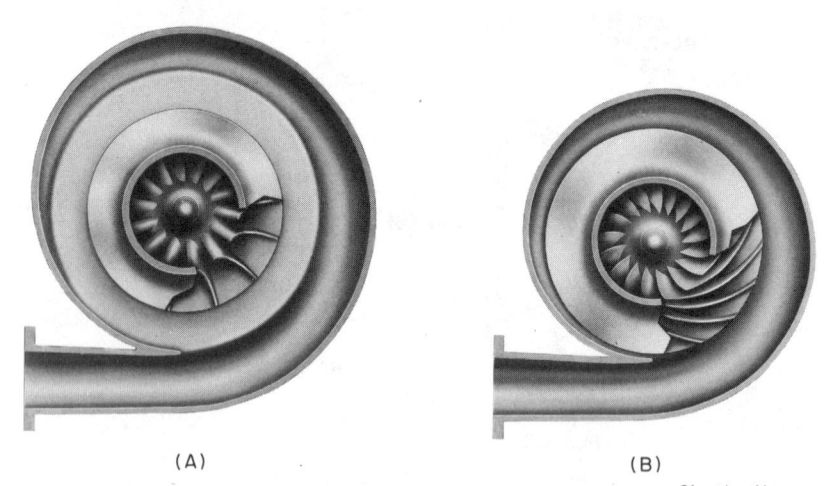

Fig. 4-22. (A) Radial-blade and (B) backward-curve centrifugal compressors.

Chrysler Airtemp

the backward-curved type. The disadvantage of radial-blade centrifugal compressors is that the energy conversion from velocity to static pressure is not efficient, and the power requirement per unit of work is, therefore, higher. Another serious disadvantage is that surge occurs at high capacity requirements, and it has a tendency to make excessive noise.

Backward-Curved Impeller Compressor

With a backward-curved impeller, the discharge gas is directly compressed by the centrifugal action of the whirling impeller and discharged into a volute, Fig. 4-22(B). The compressed gas is picked up in increasing volume by the snail-shell-shaped volute and is then discharged. This design results in a compressor that can operate quietly and efficiently over the entire capacity range.

Open and Closed Compressors

Centrifugal compressors, like reciprocating compressors, can be divided into two general types: open or closed. In general, the open-type compressor is geared to the driving mechanism and operates at speeds higher than those of the driving mechanism.

Successful operation of open-type compressors is dependent upon the shaft seal. The seal design for centrifugal machines is particularly critical, because the seal must perform several functions:

1. Seal against leakage of air into the system when the compressor is operating below atmospheric pressure

Basic Refrigeration Components

2. Seal against outward leakage of refrigerant when the compressor is operating above atmospheric pressure
3. Seal against oil entering the system
4. Perform the above functions both during operation and when the machine is shut down

A variety of seals have been developed to meet these requirements. Rotating seal rings are fixed to the shaft and rotate at compressor speed. They have highly-polished surfaces, and are perpendicular to the shaft. Stationary seal rings are attached to the compressor casing, and do not rotate. They are free to move along the axis of the compressor shaft when displaced due to thrust forces. They have highly polished surfaces which contact the rotating seal. This is the point at which the rotating element is sealed from the stationary structure of the compressor.

The seal surfaces are lubricated and cooled by oil. Oil which may leak into the compressor past the seals is returned directly to an oil reservoir, usually by gravity drainage. Oil leaking outward past the seals is collected in an oil return chamber and returned either manually or automatically to the oil reservoir.

Variations in seal design involve the physical structure and arrangement of the stationary and rotating rings, lubrication and cooling provisions, and the method of seal oil return.

Condensers

Condensers are heat exchangers designed to get rid of the heat absorbed by the refrigerant in the evaporator and the heat of compression added by the compressor. This is accomplished by condensing the hot gas discharged by the compressor and transferring the heat to some external cooling medium, usually water or air.

The diagram in Fig. 4-23 indicates what happens to a refrigerant, such as Refrigerant-12, as it passes through the condenser.

1. Point (a) represents a pressure of 136.4 psig and a temperature of 140°F. This is a SUPERHEATED gas.
2. Point (a', 110°F) represents the gas after it has been partially cooled. This point is called the SATURATED state, because further cooling results in the formation of fog, or drops of liquid.
3. Path (a-a') shows that by removing 5.59 Btu the superheated gas is brought down to the saturated condition. This process is called DE-SUPERHEATING; in other words, the reverse of superheating.
4. Path (a'-b') shows that any further removal of heat does not change the refrigerant temperature. Along this path the refrigerant is changing from a gaseous to a liquid form: in other words, the gas is being CONDENSED. At point (a') the refrigerant was entirely gas. The further left the refrigerant goes along path (a'-b'), the more liquid it becomes.

5. Point (b′) represents the refrigerant when all of it has changed to LIQUID form; its temperature here is still 110°F. Thus, in order to condense the gas and change it to a liquid, a heat removal of 54.31 Btu (87.84 − 33.53) per pound of refrigerant was required.
6. Path (b′-b) represents additional cooling of the refrigerant, which is referred to as SUBCOOLING of the refrigerant.
7. Point (b) represents the refrigerant as it leaves the condenser. The refrigerant is in liquid form and is 10°F below the saturation temperature. The heat removed by the condenser to subcool the refrigerant to this point amounted to only 2.43 Btu per pound of refrigerant.

A condenser should have sufficient surface and capacity to produce 5° to 10° subcooling of the refrigerant liquid from the time it is condensed until it leaves the condenser. Subcooling is necessary in order to insure delivery of liquid to the expansion device. If there is no subcooling, a slight warm-up in the liquid line can produce a flash-back to gas, and thereby degrade system performance.

Refrigeration condensers are divided into two basic types; air-cooled and water-cooled as shown in Fig. 4-24. Both types work on the principle of increasing the sensible heat content of the cooling medium. The cooling medium is increased in temperature by a certain number of degrees; enough to absorb the heat taken from the condensing refrigerant.

Air-Cooled Condensers

Air cooling was first used on small condensing units for household refrigerators and room air conditioners, but is now common in central air conditioning systems as well.

Air-cooled condensers, Fig. 4-25, are generally made as continuous tube single-row or continuous tube double-row radiator types for small systems. Large systems are provided with multiple-row radiator type condensers. The hot refrigerant gas flows from the compressor into the top of the condenser. As previously mentioned, the gas arrives at this point in a superheated condition. Air passing over the fins and tubes of the condenser first removes the superheat from the refrigerant and then condenses the gas into a liquid. This liquid flows out the bottom of the condenser (to a liquid receiver, if used) to the metering device. A condenser fan provides a steady flow of air for removal of heat from the finned tubing and consequently from the refrigerant that flows through the finned tubes. The air-cooled condenser requires large amounts of air for efficient condenser cooling.

Air-cooled condensers are classified as chassis-mounted, i.e., a component of a compressor-condensing unit; or as remote, an integral assembly remote from the compressor. Both of these are further subdivided into indoor and outdoor type units.

Basic Refrigeration Components

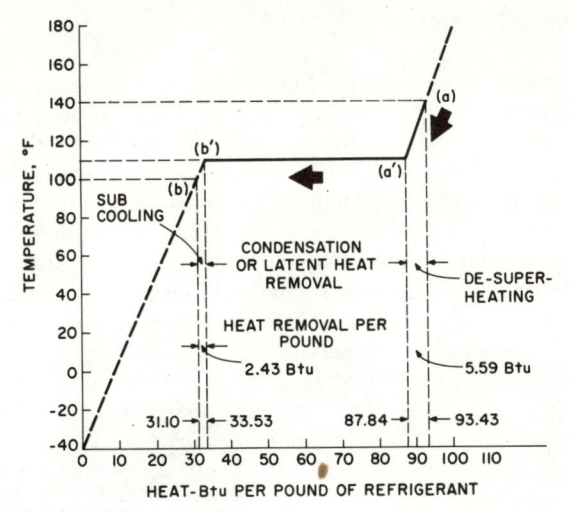

1-63

Fig. 4-23. Heat transfer of Refrigerant-12 in a condenser.

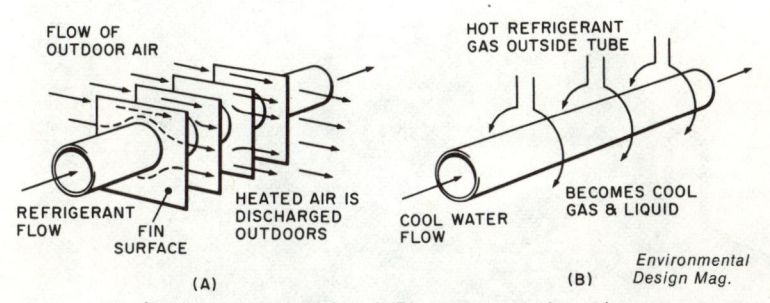

Fig. 4-24. (A) Air-cooled and (B) water-cooled condensers.

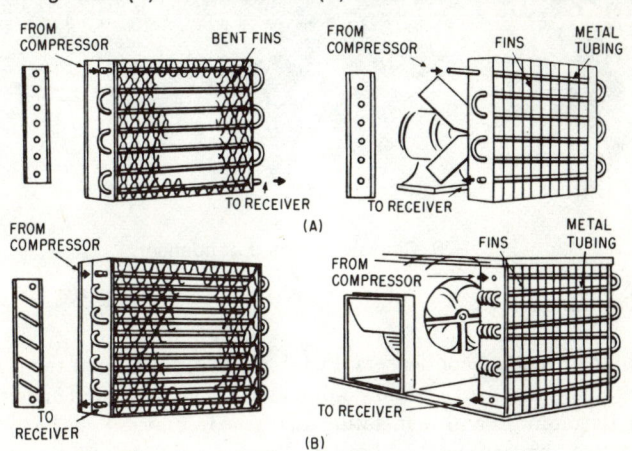

Environmental Design Mag.

Fig. 4-25. Basic air-cooled condensers: (A) continuous tube, radiator, and (B) multiple-row radiator types.

Chassis-Mounted Condensers

Chassis-mounted condensers, Fig. 4-26, are usually mounted on a common base with the compressor and motor. The condenser fan may be mounted on the shaft of the compressor motor. Chassis mounted air-cooled condensers are seldom used for indoor units of over 5 horsepower because of their physical size. Outdoor units obviously do not have this type of restriction, and also have the advantage of an abundant supply of clean, comparatively cool air. Outdoor units include separate condenser fans driven by their own motors so that all compressor motor power is utilized to produce refrigeration. The entire assembly is enclosed in a casing which is removable for servicing.

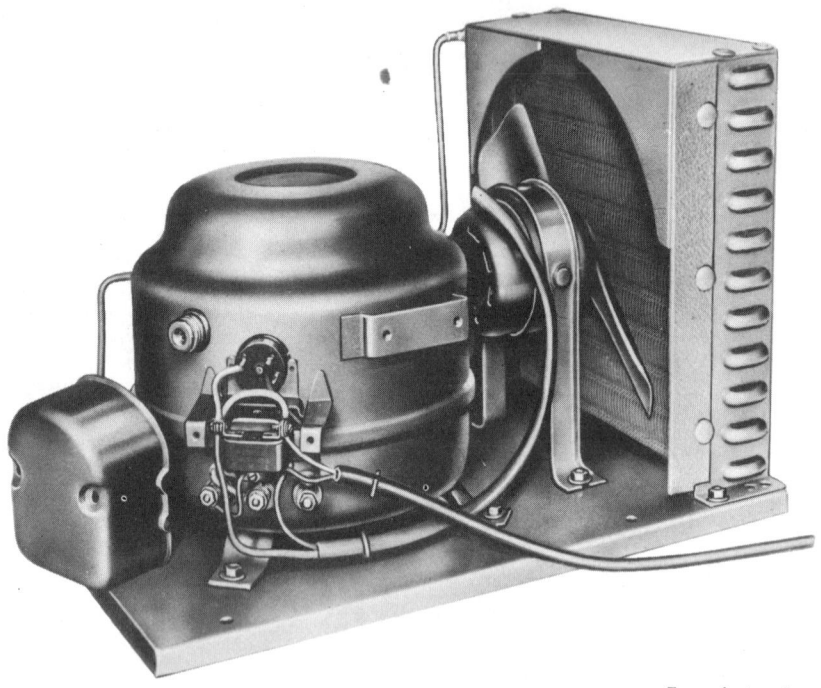

Texas Instru. Co.
Fig. 4-26. Chassis-mounted condenser.

Remote Condensers

Remote air-cooled condensers, Fig. 4-27, are generally of the forced-air type. They consist of three main components: a finned condensing coil, a fan (with its motor), and a metallic casing. Air may either be drawn or blown through the coil by the fan. In the draw-through unit, the fan is located on the air discharge side of the coil. On a blow-through type, the motor and fan are located in the entering air stream. Remote gravity type units consist of a single-row condensing coil of large face area, usually mounted on a roof at a 7° angle with the horizontal. Since air movement through the coil takes place due to gravity or wind, the condenser capacity is highly sensitive to wind velocity. Selection of the gravity type unit is made on the basis of expected minimum wind velocity, and is recommended where ample roof or floor surface is available, fan noise must be completely avoided, and power cost is high enough so that the saving in motor elimination becomes significant.

Dunham-Bush Inc.
Fig. 4-27. Remote air-cooled condenser.

The principal disadvantage of the air-cooled condenser is the power required to move the air and the reduction of capacity on hot days. This loss of capacity, due to the high condensing pressures needed on hot days, requires that equipment of increased capacity be selected to meet peak load. The principal advantages are simplicity and low installation costs; hence, they are frequently used in small self-contained units.

Water-Cooled Condensers

Water-cooled condensers are commonly used with compressors of one horsepower or larger. They usually prove to be the most economical choice if a cheap and adequate water supply, and means for its disposal, are available. Water is not as subject as air to extreme variations in temperature and, most important, its temperature is usually well below the temperature of the outside air during the maximum cooling load period (summer). Therefore, a system with a water-cooled condenser retains its cooling capacity well on extremely hot days.

Water-cooled condensers for residential air conditioning are made in many forms, the most common being: shell-type and tube-within-tube type.

Shell-Type Condensers

Shell-type water-cooled condensers are widely used and have the advantage of being compact in size, and serve the dual function of condenser and receiver. The shell-type condenser is subdivided into two basic designs: shell-and-tube condenser, and the shell-and-coil condenser.

The shell and tube water-cooled condenser, Fig. 4-28(A), has tubes running between end sheets that are fastened in a shell having removable heads to enable tube cleaning or replacement. Cooling water enters through the heads which are baffled to make the water make one or more passes through the tubes.

The shell-and-coil condenser, Fig. 4-28(B), is a modification of the shell-and-tube type, and is often used on small packaged units. In the shell-and-coil condenser, water passes through a continuous spiral or hairpin-wound coil which may be divided into several parallel sections. The shell ends are not removable, but water scale can be removed by treatment with the proper acid solution.

Well-designed shell-type condensers have the water coils arranged in free-draining arrangement to assist in draining off the water when the system is not in use. Many also have external fins on the tubes, Fig. 4-28(A), for better heat transfer.

In the shell-type condenser, the tubing is arranged in the shell so that a small portion, about 10%, of its surface is submerged below the normal refrigerant level. The length of the water pass is very small, but it is still necessary to go through the steps of de-superheating, condensing and subcooling. The de-superheating is done very quickly in the upper portion of the shell. In the middle portion of the shell, the gas is turned to liquid, and the condensed liquid then drips to the bottom and is subcooled by the submerged portion of the tubing.

Tube-Within-Tube Condensers

Tube-within-tube condensers are fabricated with one or more water tubes inside a refrigerant tube. The simpler types are usually wound in the shape of trombone coils. The advantages of counterflow between water and refrigerant are obtained, with the cooling water entering at the bottom of the condenser, and the hot gas entering at the top and passing down-

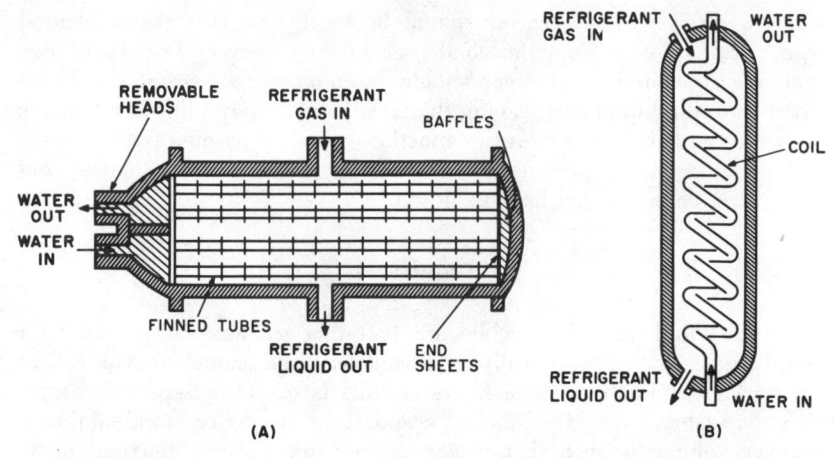

Fig. 4-28. Water-cooled condensers: (A) shell-and-tube and (B) shell-and-coil types.

ward between the two tubes. The incoming water readily subcools the liquid refrigerant leaving at the bottom, and the efficiency of heat transfer is very good.

Trombone-shaped condensers are usually limited to about 3 horsepower in size because of the problems of bending larger tubes. Mechanical (brush) cleaning is not possible and chemical cleaning is necessary.

Tube-within-tube brush-cleanable condensers, Fig. 4-29, are built in a wider range of sizes than trombone-shaped types, with straight tubes between headers. Removable header plates give access to the inside of the water tubes for mechanical cleaning. Individual banks of tubes are combined to give greater capacity. Tube-within-tube condensers require a

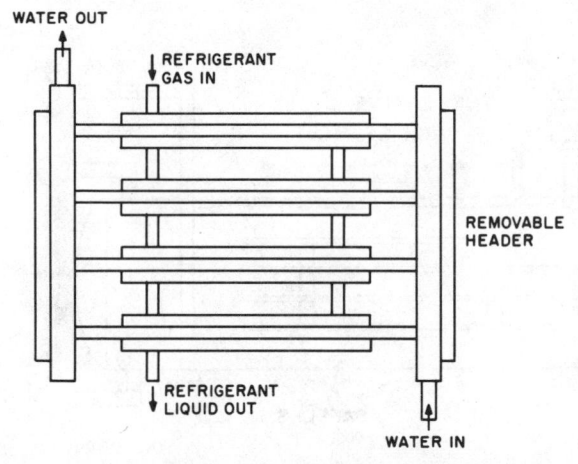

Fig. 4-29. Tube-within-tube condenser.

separate receiver. The receiver should be located so that the condensed liquid drains freely from the condenser to the receiver. The liquid line from the condenser to receiver should be ample in size, and should be larger than the liquid line leaving the receiver. The large line between the condenser and receiver serves two functions. First, it permits free drainage of liquid; second, it permits any gas in the receiver to rise to the condenser and be condensed back to liquid.

Receivers

The liquid receiver is a collecting tank that acts as a storage point for liquid refrigerant. Theoretically, it should be large enough to hold all of the refrigerant in the system. However, this is not always possible, especially on systems with large flooded evaporators. In any case, the minimum receiver volume is such that it can absorb the seasonal fluctuations in liquid refrigerant requirements, and prevent build up into the condenser during operation. Although the receiver is a part of the high side, its size is determined by the design and operation of the low side.

Air-cooled condensers have a smaller refrigerant storage volume and, on systems where storage for the entire refrigerant charge is necessary, they require a separate receiver. Receivers also provide storage for the extra refrigerant charge for systems with head pressure (discharge pressure) control that require flooding of the condenser coil to reduce capacity during winter operation. On systems that vary the amount of air flowing across the condenser, thus controlling head pressure, a receiver is not required.

Liquid receivers are generally classified as horizontal or vertical. Separation and removal of lubricating oil from the refrigerant in horizontal receivers is accomplished by the use of baffles, Fig. 4-30. These baffles

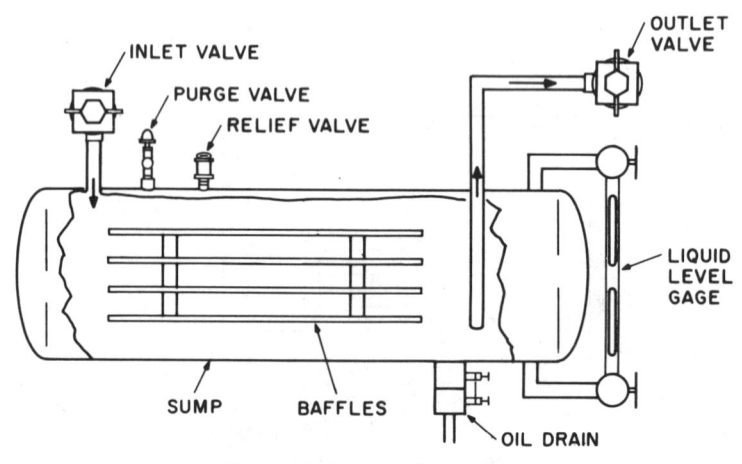

Fig. 4-30. Refrigerant receiver.

tend to quell agitation of the incoming liquid refrigerant, so that the oil separates readily. The receiver is tilted slightly so that oil will drain to the sump, where it can be periodically removed. Receivers are generally provided with level gauges to indicate the level of liquid refrigerant.

All receivers have liquid inlet and outlet connections. Other connections which are normally furnished include: relief valves, purge valves, liquid level gauges, and oil drain valves.

Refrigerant Metering Devices

The major control problem in modern air conditioning systems is the method of expanding a high-pressure liquid refrigerant into a low-pressure, wet-gas mixture in exact proportion to the rate of evaporation in the evaporator. Modern systems use a variety of devices to control the flow of refrigerant: capillary tubing, pressure-operated valves, thermostatically-operated valves, float valves, and orifices (distributors).

Capillary Tubes

A capillary tube is sometimes used in place of a thermostatic expansion valve to restrict or meter the flow of refrigerant into the evaporator. With the advent of the hermetic compressor and the halocarbon refrigerants, the capillary tube became practicable and rapidly achieved popularity, especially with smaller hermetic units such as room air conditioners.

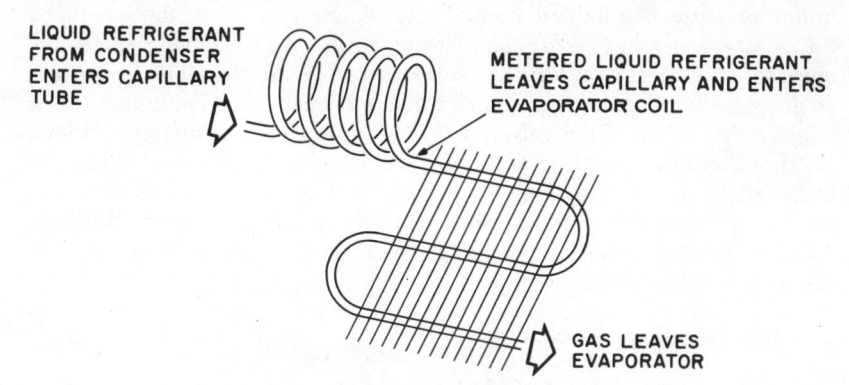

Fig. 4-31. Capillary tube operation.

The capillary tube operates on the principle that liquids pass through it much more readily than gases. It consists of a small diameter line which connects the condenser outlet to the evaporator inlet. It is sometimes soldered to the outer surface of the suction line for heat exchange purposes.

During compressor operation, Fig. 4-31, the capillary tube reduces the pressure of the refrigerant from high-side pressure to low-side pressure. When the compressor stops, pressures equalize in the system; that is, the liquid refrigerant in the condenser continues to flow into the evaporator until the two pressures are approximately equal. During this situation, a large share of the refrigerant liquid will collect in the evaporator during the compressor shutdown period. To prevent liquid refrigerant from reaching the compressor, an ACCUMULATOR (fluid-storage device) may be added to the system. However, most of the systems utilizing a capillary tube are designed to incorporate an evaporator large enough to hold the entire liquid charge. Inherently, a capillary tube does not operate as efficiently over a wide range of conditions as does a thermostatic-expansion valve, but its performance is generally very good.

Constant-Pressure Expansion Valve

One of the earliest means of refrigerant control was the constant-pressure (automatic) expansion valve, which was developed to eliminate the problems encountered with hand-operated valves. The main objection to the latter was that, as the load changed, the setting on the valve had to be changed, either to prevent flooding the compressor with liquid refrigerant, or to prevent starving the evaporator.

The constant-pressure expansion valve, Fig. 4-32, was designed to operate on evaporator pressure and keep it constant, since the refrigerant pressure in the evaporator determined the evaporator temperature.

Its basic construction consists of an adjustable spring on top of a diaphragm which exerts its force in an opening direction, and a spring beneath the diaphragm which exerts its force in a closing direction. Evaporator pressure is admitted beneath the diaphragm, hence the combined forces of evaporator pressure and closing spring act to counterbalance the opening spring pressure. With the valve set and feeding at a given pressure, an increase in evaporator pressure will act beneath the diaphragm, force it upward, and cause the valve pin to move in a closing direction, which restricts the flow of refrigerant and limits evaporator pressure. When the evaporator pressure, because of a change in load, drops below the valve setting, the top spring pressure moves the valve pin in an opening direction, which increases the refrigerant flow, and raises the evaporator pressure to the proper valve setting.

Thermostatic-Expansion Valve

As the demands of the refrigeration industry became more complex, the constant-pressure expansion valve failed to meet the requirements. It became apparent that a valve that could modulate to meet the variations in load and not starve the evaporator or permit liquid flooding the compressor had to be developed. This led to the development of the thermostatic-expansion valve, which is a precision device designed to regulate the rate of liquid refrigerant flow into an evaporator in exact proportion to the rate of evaporation of the liquid refrigerant. The thermostatic-expansion valve responds to both the temperature of the refrigerant gas leaving the evaporator, and the pressure in the evaporator.

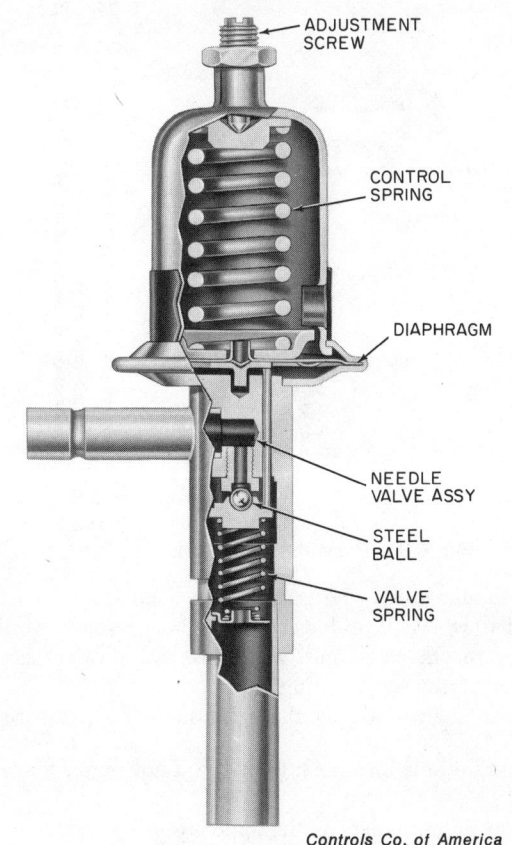

Fig. 4-32. Adjustable constant-pressure expansion valve.

Construction of the thermostatic-expansion valve, Fig. 4-33, is similar to that of the constant-pressure expansion valve, but also incorporates a power element responsive to changes in the DEGREE OF SUPERHEAT of the refrigerant gas leaving the evaporator. This power element consists of a bellows connected by means of tubing to a thermal feeler bulb fastened to the suction line from the evaporator, Fig. 4-34. The bulb, bellows, and tube are usually charged with the same liquid refrigerant used in the evaporator.

Installation of the thermostatic-expansion valve and proper location of the bulb are fully as important as correct valve selection. The valve should always be installed as near the liquid header as possible. An upright position with the power assembly on top is preferred. Since practically all thermal expansion valves used in air conditioning are of the LIQUID-CHARGE type, the valve may be installed in either warm or cold locations. If the valve is of the GAS-CHARGE type, the valve must be located to insure the valve body being warmer than the thermal bulb at all times.

1-72   Basic Air Conditioning

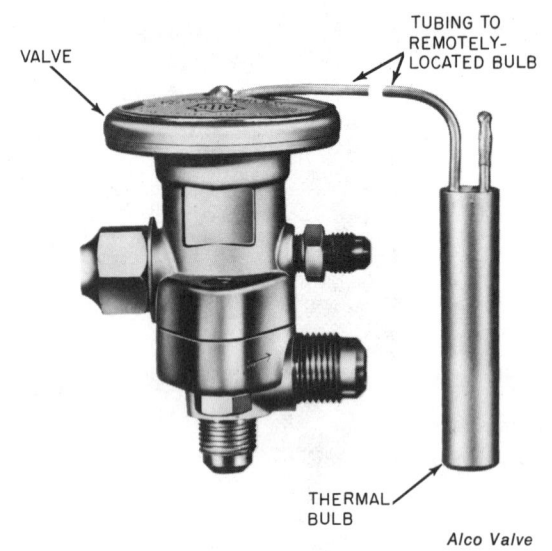

Fig. 4-33. Thermostatic expansion valve.

    The bulb should always be installed on a horizontal run, and should never be installed where liquid can trap. The presence of liquid in the suction line near the thermal bulb will cause false valve response due to the cooling effect of the evaporating liquid.

    Operation is determined by three fundamental pressures, Fig. 4-35:

P_1: Bulb pressure acts on one side of the diaphragm, tends to open the valve
P_2: Evaporator pressure acts on the opposite side, tends to close the valve
P_3: The pressure of the superheat spring, assists in the closing action

    When the valve is feeding liquid refrigerant, bulb pressure is balanced by the evaporator pressure and spring pressure:

$$P_1 = P_2 + P_3$$

    With the same refrigerant used in both the thermostatic element (bulb) and the cooling system, each exerts the same pressure when their temperatures are identical. After evaporation of the liquid refrigerant in the evaporator the suction gas is superheated. However, the evaporator pressure, neglecting pressure drop through the evaporator, is unchanged. The warmer vapor flowing through the suction lines increases the bulb temperature. Since the bulb contains both vapor and liquid refrigerant (not superheated vapor alone, as in the suction line) its temperature and pressure increases. This higher bulb pressure acting on the top side (bulb side) of the diaphragm is greater than the pressure exerted by the opposing evaporator and spring, which causes the valve pin to be moved away

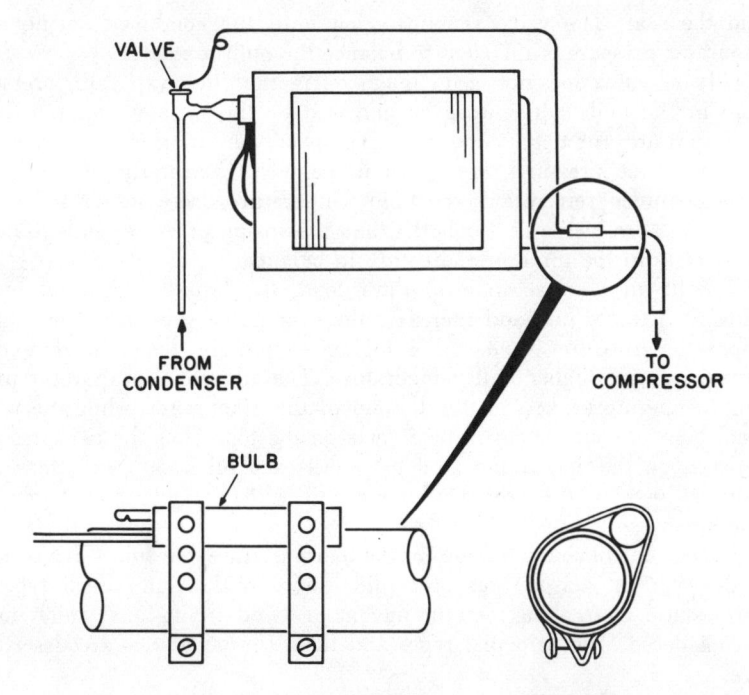

Fig. 4-34. Thermal bulb location on suction line.

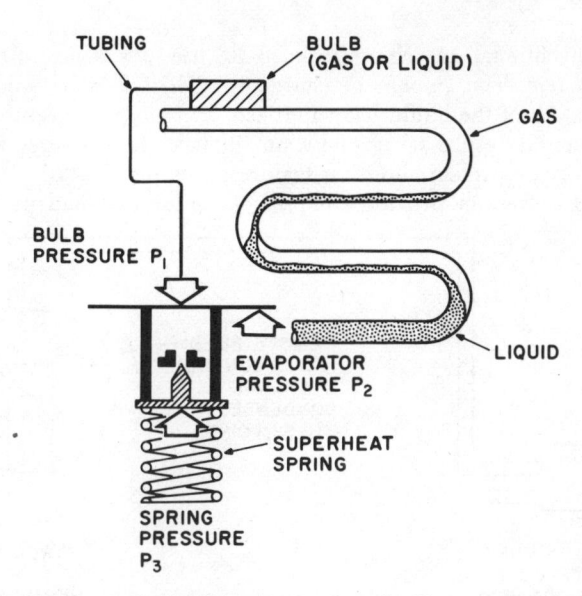

Fig. 4-35. Thermostatic expansion valve showing pressures acting on diaphragm.

1-74   Basic Air Conditioning

from the seat. The valve remains open until the combined spring and evaporator pressure is sufficient to balance the bulb pressure.

If the valve does not feed enough refrigerant, the evaporator pressure drops or the bulb temperature is increased by the warmer vapor leaving the evaporator (or both), and the valve opens, admitting more refrigerant until the three pressures are again in balance. Conversely, if the valve feeds too much refrigerant, the bulb temperature decreases or the evaporator pressure increases (or both), and the spring pressure tends to close the valve until the three pressures are in balance.

With an increase in evaporator load, the liquid refrigerant evaporates at a faster rate and increases the evaporator pressure. The higher evaporator pressure results in a higher evaporator temperature and a correspondingly higher bulb temperature. The additional evaporator pressure (temperature) acts on the bottom of the diaphragm, while the additional bulb pressure (temperature) acts on the top. Thus the two pressure increases on the diaphragm tend to cancel each other out, and the valve easily adjusts to the new load condition with a negligible change in superheat.

The control characteristics of the thermostatic-expansion valve depend on the type of charge (gas or liquid) used in the bulb. Each type of thermostatic charge has certain advantages and limitations which must be considered. The principal types and their characteristics are described here.

Gas Charge

A conventional gas charge, Fig. 4-36, uses the same refrigerant as does the system. The amount of charge is such that, at a predetermined temperature, all of the liquid has vaporized, and any temperature increases above this point results in virtually no increase in pressure. Because of the characteristic pressure-limiting feature of its thermostatic element, the gas-charged valve can provide compressor motor overload protection on

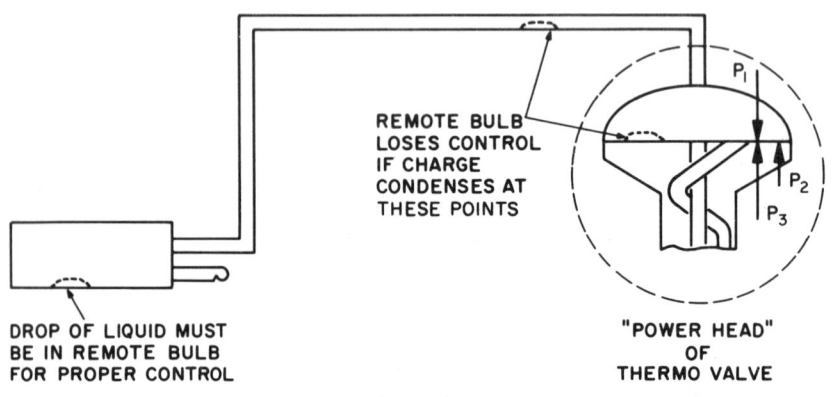

Fig. 4-36. Gas-charged expansion valve.

some systems by limiting the maximum operation suction pressure (MOP). It also helps to prevent flooding-back on starting. It is normally used in the 30–50°F evaporator temperature range and will not operate in cross ambient conditions.

Liquid Charge

Conventional liquid charges also employ the same refrigerant in the thermostatic element as in the cooling system. They are termed cross-ambient charges because the volumes of the bulb, capillary tubing, and diaphragm chamber are proportioned so that the bulb contains some liquid under all temperature conditions. Therefore, the bulb will always control valve operation even with a colder diaphragm chamber or capillary tube.

The characteristics of the liquid-charged valve, Fig. 4-37, result in an increase in operating superheat as the evaporator temperature decreases; that is, the valve setting required for a reasonable operating superheat at a low evaporator temperature may cause flooding back during pulldown from ambient temperature (initial cooling period). This usually limits its use to moderately-high evaporator temperatures.

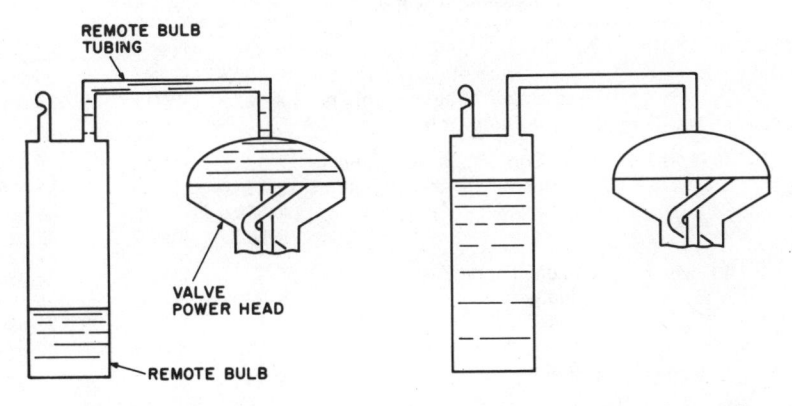

Fig. 4-37. Liquid-charged expansion valve. Since the volume of liquid charge exceeds the combined volume of the power head and remote bulb tubing, there is always some liquid in the remote bulb.

Liquid Cross Charge

Liquid cross charges are also cross-ambient charges, but unlike conventional liquid charges, they employ a liquid in the thermostatic element (bulb) which is different from the refrigerant in the system. Cross charges have more linear pressure-temperature characteristics than do the system refrigerants with which they are used; that is, cross charges in the commercial temperature range generally have superheat characteristics which are nearly constant, or which deviate only moderately through the evaporator temperature range. Thus, flooding-back problems are minimized.

All-Purpose Charge

Through research and engineering development, the manufacturers of thermostatic-expansion valves have perfected what can be called an all-purpose charge. This charge combines the best properties of the standard liquid and gas charges, and is not limited in temperature range or type of application by any inherent properties.

As shown in Table 4-1, the all-purpose charge provides superior performance for all applications with evaporator temperatures ranging from 50°F to −40°F, the range previously covered by all other charges combined. It has nearly constant superheat and, therefore, does not tend to cause flooding back at either start-up or at minimum evaporator temperature. Its flexibility permits accurate operation on practically all types of systems.

Table 4-1. All-Purpose Charge Application Data Pressure Limiting (M.O.P.) Required for Compressor Motor Overload Protection

Evaporator Temperature Range	Application Examples	System Refrigerant	Charge Symbol	Standard M.O.P.
+50°F to −40°F	Air Conditioning, Unit Coolers, Cases, Industrial Processing, Water Chillers, Heat Pumps, etc.	R-12 R-500 R-502	FW55 CW65 RW110	55 psig 65 psig 110 psig
+20°F to −40°F	Unit Coolers, Brine Chillers, Industrial Processing, etc.	R-12 R-22 R-500 R-502	FW35 HW65 CW45 RW75	35 psig 65 psig 45 psig 75 psig
0°F to −40°F	Frozen Food Cases, Locker Plants, etc.	R-12 R-22 R-500 R-502	FW15 HW35 CW20 RW45	15 psig 35 psig 20 psig 45 psig

Equalizers

Equalizers are devices used with thermostatic valves to equalize the pressures on both sides of the valve diaphragm.

On some evaporators, particularly the larger ones, there is an appreciable pressure drop between the valve outlet and the bulb location. A thermostatic-expansion valve equipped with an INTERNAL EQUALIZER, Fig. 4-38, can seriously reduce the evaporator capacity due to excessive evaporator pressure drop which causes the valve to operate at abnormally high superheat.

Figure 4-39 shows a typical system using an internal equalizer and with no pressure drop between valve and bulb. The pressure at the valve outlet and at the bulb location is 27 psi. The evaporator side of the valve diaphragm senses the evaporator pressure of 27 psi, plus the spring pres-

Basic Refrigeration Components

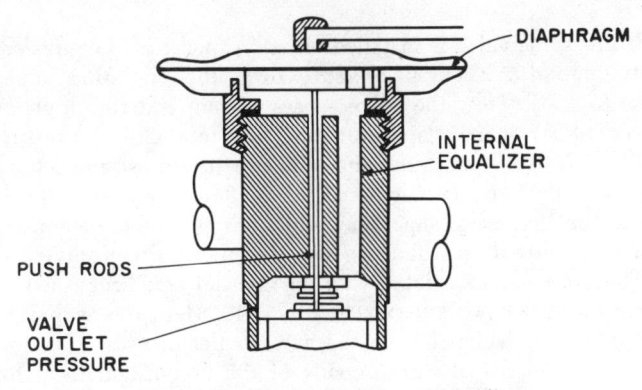

Fig. 4-38. Valve with internal equalizer.

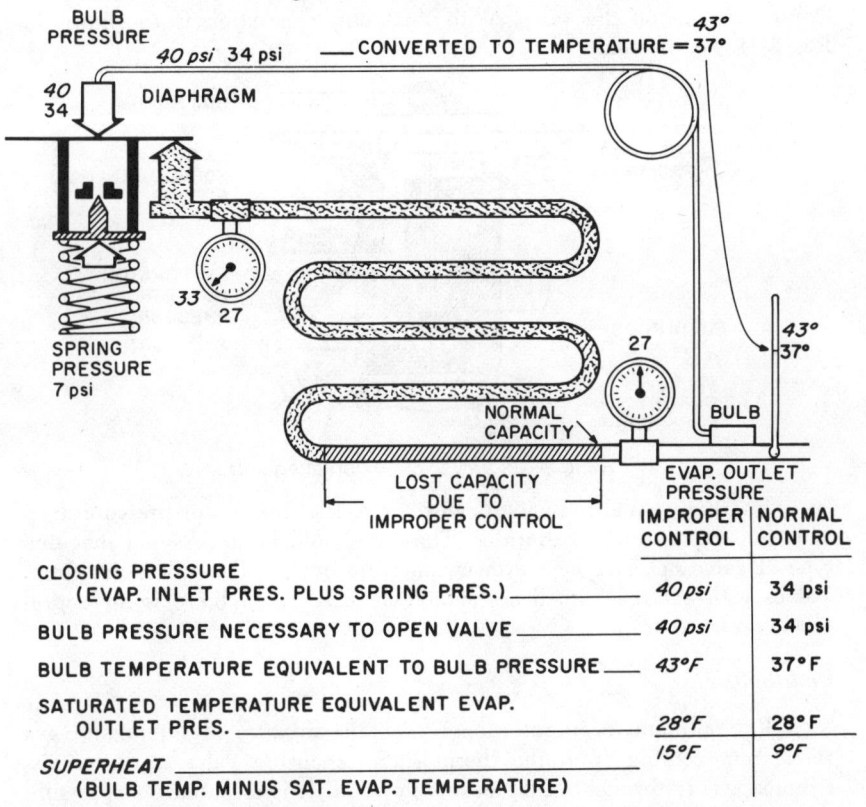

Fig. 4-39. Normal and improper valve control with internal equalizer.

	IMPROPER CONTROL	NORMAL CONTROL
CLOSING PRESSURE (EVAP. INLET PRES. PLUS SPRING PRES.)	40 psi	34 psi
BULB PRESSURE NECESSARY TO OPEN VALVE	40 psi	34 psi
BULB TEMPERATURE EQUIVALENT TO BULB PRESSURE	43°F	37°F
SATURATED TEMPERATURE EQUIVALENT EVAP. OUTLET PRES.	28°F	28°F
SUPERHEAT (BULB TEMP. MINUS SAT. EVAP. TEMPERATURE)	15°F	9°F

sure of 7 psi, a total of 34 psi valve-closing pressure. The valve, consequently, adjusts its flow rate until the suction line vapor becomes sufficiently superheated to create a bulb temperature of 37°F, which develops a pressure of 34 psig, balancing the evaporator and spring. The resulting superheat is 9°F.

If this same valve is installed on an evaporator of equivalent nominal capacity but with a 6 psi pressure drop the operating superheat will increase to 15°F. Now the valve senses a comparatively high pressure of 33 psi (27 + 6) at the evaporator inlet. The total closing pressure is therefore, 33 + 7, or 40 psi. Since the bulb pressure must equal the total closing pressure, the valve now must reduce its flow rate even more than before to create the necessary superheat and bulb pressure, demonstrating that excessive pressure drop will cause loss of evaporator capacity.

The problems associated with an internal equalizer can be alleviated by using an EXTERNAL EQUALIZER, Fig. 4-40. Here the evaporator side of the diaphragm is isolated from the valve outlet pressure. Suction pressure is transmitted to the evaporator side of the diaphragm by a line usually connected between the suction line near the evaporator outlet and an external fitting on the valve. With this setup, operation is as shown in Fig. 4-41, i.e., normal.

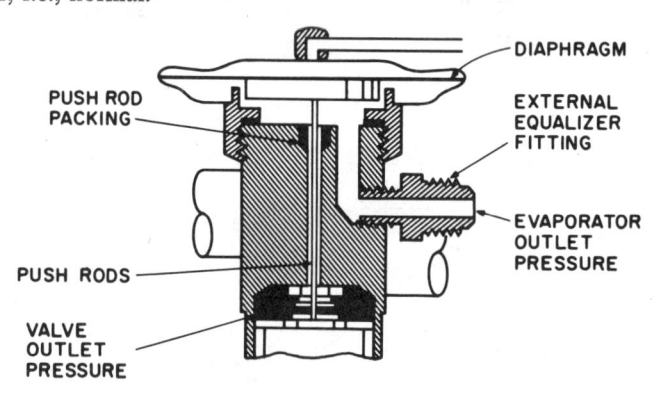

Fig. 4-40. Valve with external equalizer.

Internally-equalized valves can tolerate less evaporator pressure drop at lower evaporator temperatures. Thus, it should be understood that this type of valve operates best with evaporators having a low pressure drop. Valves with external equalizers should be used when there is an appreciable pressure drop.

Distributors

For optimum evaporator operation, the mixture of liquid and gas refrigerant coming from the thermostatic-expansion valve must be distributed evenly throughout the evaporator coil. It is essential that pressure drop within the coil be held within reasonable limits, and that refrigerant flow be such as to ensure proper return of oil to the compressor. In all but the smallest evaporator coils, these requirements are met by incorporating multiple refrigerant circuits. For certain applications, each circuit can be served by its own thermostatic-expansion valve, but normal practice is to employ a distributor that evenly proportions refrigerant flow from the valve to all the circuits.

BULB PRESSURE 34 psi — CONVERTED TO TEMPERATURE = 37°

DIAPHRAGM

SUCTION PRESSURE AT BULB = 27 psi

EVAPORATOR INLET PRESSURE 33 psi

SPRING PRESSURE 7 psi

EVAPORATOR OUTLET PRESSURE 27 psi

CLOSING PRESSURE _____ = 27 + 7 = 34 psi
(SUCTION PRESSURE AT BULB PLUS SPRING PRESSURE)
BULB PRESSURE NECESSARY TO OPEN VALVE _____ 34 psi
BULB TEMPERATURE EQUIVALENT TO 34 psi _____ 37° F
SATURATED TEMPERATURE EQUIVALENT TO EVAPORATOR
    OUTLET PRESSURE _____ 28° F
SUPERHEAT _____ 9° F
    BULB TEMPERATURE MINUS SATURATED EVAPORATOR
    TEMPERATURE

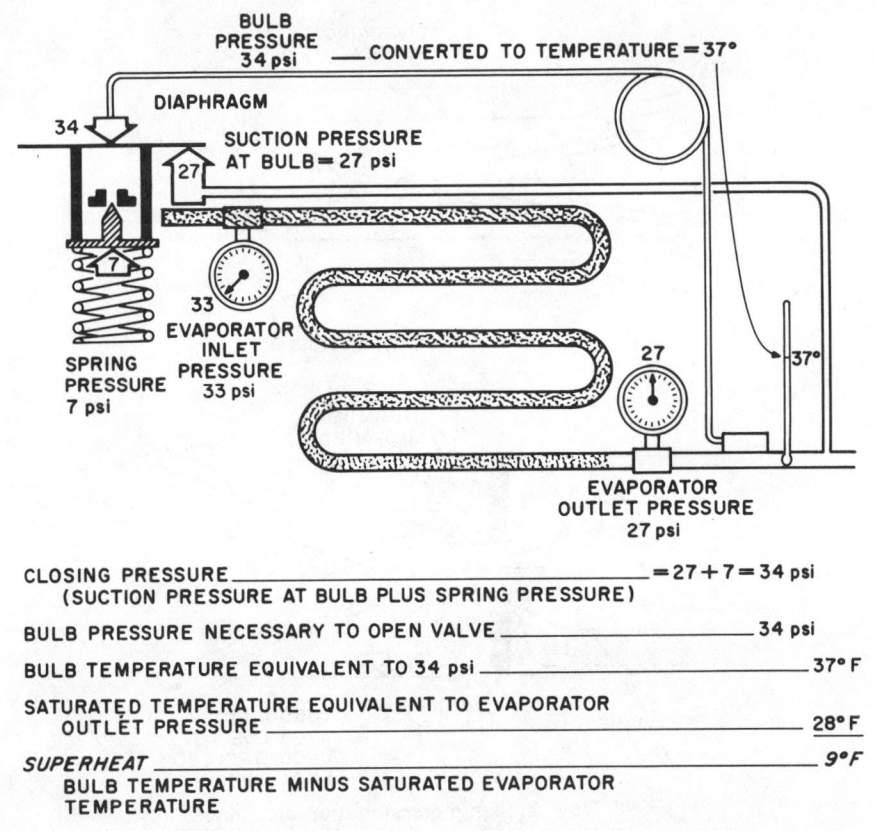

Fig. 4-41. Proper valve control with external equalizer.

From the strictly functional point of view, a refrigerant distributor should not only provide even distribution, but should do so with a minimum of pressure drop to assure stable control by the thermostatic-expansion valve.

One of the more recent developments in flow distribution is the VENTURI-FLO DISTRIBUTOR, Fig. 4-42. This distributor consists of a converging section, a throat, and a diverging section. The smooth contoured approach to the converging section prevents turbulence in the refrigerant flow during the transition between the approach and the throat. The diverging section following the throat reduces, through expansion, the high throat velocity of the refrigerant, converting the velocity energy back to pressure energy. Because of the venturi design, pressure drop through the distributor is minimal, and is due only to wall friction losses. The refrigerant flow pattern is smooth and assures even distribution to all tubes feeding the individual circuits of the coil.

Another type of distributor, with a different operating principle, is the PRESSURE-DROP DISTRIBUTOR, Fig. 4-43. The pressure-drop distributor consists of a straight approach to an orifice plate which contains a con-

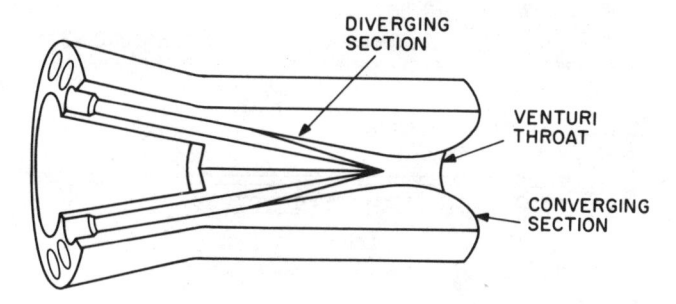

Fig. 4-42. Venturi-Flo ® type distributor.

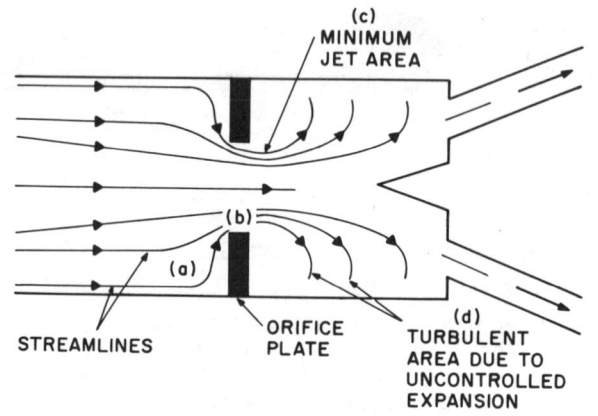

Fig. 4-43. Flow through a pressure-drop distributor.

centric hole. Operation depends on the high pressure drop across the orifice, and the resulting turbulence, to provide distribution.

In this type of distributor, the approaching refrigerant stream lines nearest the pipe walls (Point A of Fig. 4-43), turn inward rather abruptly at the upstream face of the plate and flow parallel to the plate toward the orifice. This flow continues past the edge of the orifice and is swept along through Point B of the orifice by the center mass of the refrigerant flow. The refrigerant flow continues through the orifice, with the minimum jet area (Point C) forming downstream of the orifice plate. The uncontrolled expansion in the large downstream section causes a turbulent pattern to be set up between the minimum jet area, the orifice plate, and the walls (Point D). The velocity energy is not converted to pressure energy, but is dissipated in violent turbulence in the downstream section.

Another type of refrigerant distribution device in use is the CENTRIFUGAL DISTRIBUTOR, Fig. 4-44. The centrifugal distributor depends on a high entrance velocity to swirl the liquid refrigerant around to the outlet tubes.

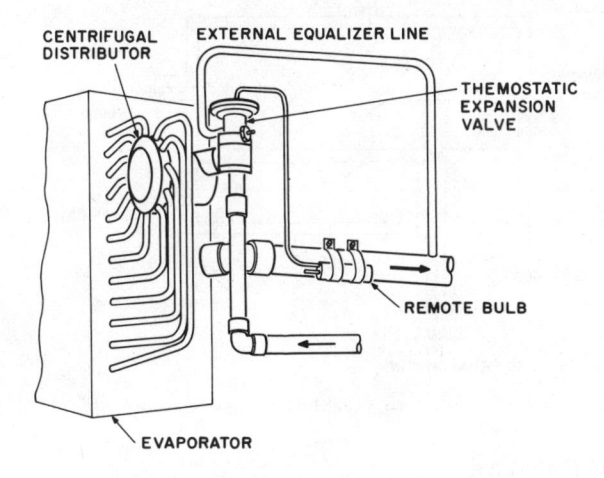

Fig. 4-44. Single outlet thermal expansion valve and centrifugal distributor.

Float Valves

The application of thermostatic expansion valves and capillary tubes to FLOODED evaporators is rather uncommon. Superheat is necessary for proper valve control, but only a few degrees of suction vapor superheat in a flooded evaporator will incur a substantial loss in capacity. The most common method of control for flooded evaporators is to use either high-side or low-side float valves.

High-Side Float Valve

The high-pressure float valve, Fig. 4-45, is an expansion valve that is operated by the liquid refrigerant level on the high-pressure side of the system. The liquid refrigerant flows from the condenser into the float valve which opens, expanding the liquid refrigerant into the evaporator. The high-pressure float valve permits refrigerant flow into the evaporator at the same rate that the refrigerant gas is pumped out of the evaporator by the compressor. The valve consists of a float chamber, a valve pin and seat, and a float ball which operates the valve pin. The arrangement of the valve pin and the float arm pivot are such that the weight of the float ball will move the valve pin in a closing direction when the liquid refrigerant level recedes in the float chamber.

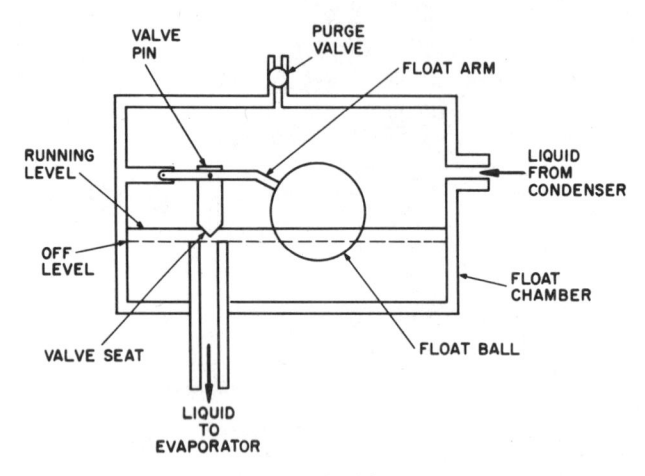

Fig. 4-45. High-side float valve.

Low-Side Float Valve

The low-side float valve, Fig. 4-46, performs the same function as the high-side float valve, but is connected to the low-pressure side of the system. The valve consists of a needle valve operated through a simple lever mechanism attached to a float ball. The valve is so mounted, with respect to the liquid level in the evaporator, that any drop in liquid level occurs in both simultaneously. This causes the float ball to drop and the needle to displace from its seat and open the valve. Additional liquid refrigerant then flows into the evaporator until the rising float ball finally closes the valve, at which time the proper operating level is re-established.

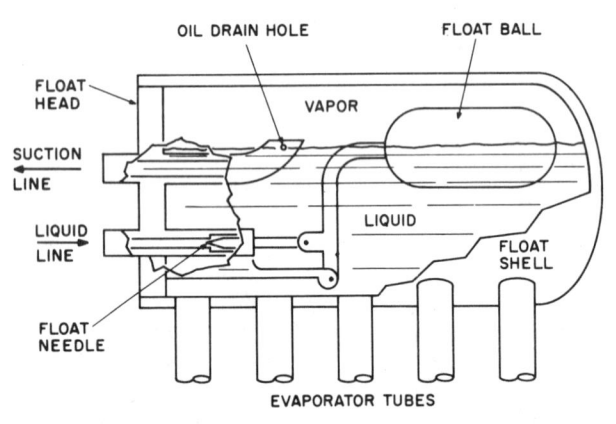

Fig. 4-46. Low-side float valve.

5
Peripheral Devices

Pressure Valves

In an air conditioning system, pressure must be controlled for several reasons:

1. To prevent excessive pressures from damaging the equipment
2. To hold the evaporator at a particular temperature, or prevent the evaporator from going below a particular temperature. For example, to prevent evaporator coil freeze-up
3. To maintain condenser (head) pressure at an efficient level of operation, or prevent head pressure from falling too low

There are many ways of controlling pressure to meet the needs which we have considered. However, there are basically three types of valves that are generally used in well designed air conditioning systems. They are evaporator pressure regulators, suction or crankcase pressure regulators, and condenser or head pressure regulators.

Evaporator Pressure Regulator

The evaporator pressure regulator, Fig. 5-1, is a device sensitive mainly to its own INLET pressure. It closes on pressure decline and opens on pressure rise. In operation, evaporator pressure is admitted beneath the diaphragm or bellows of the valve through an internal equalizer passage. When the evaporator pressure rises above the force exerted by the pressure spring, the valve moves to the open position. When the pressure drops below the force exerted by the spring, the valve closes. By setting the valve for a pressure which corresponds to a refrigerant saturation temperature above the freezing point of water, for example, ice is prevented from forming on the coil.

Suction or Crankcase Pressure Regulator

On some installations, the load may exceed the compressor motor ratings. In order to protect the motor from overloads and possible burnout as a result of these excessive suction pressures, a suction pressure regulator is installed to limit the suction pressure at the compressor. This

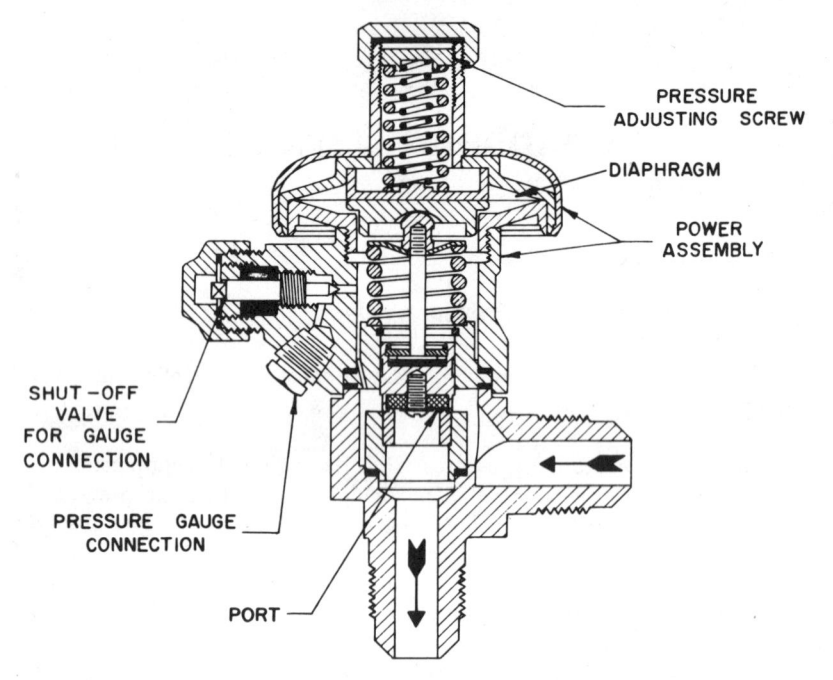

Fig. 5-1. Direct acting evaporator pressure regulator.
Alco Valve

so-called suction or crankcase pressure regulator is very similar in operation to the evaporator pressure regulator. However, this device is sensitive mainly to its own OUTLET pressure, whereas the evaporator pressure regulator is sensitive to inlet pressure.

The suction pressure regulator should be used on any installation where compressor motor protection is required because of:

1. High starting loads
2. Surges in suction pressure
3. High suction pressure caused by hot gas defrost
4. Prolonged operation at excessive suction pressures
5. Low voltage and high suction pressure conditions

Condenser Pressure Regulators

Air-Cooled Condenser

The need for some type of control arises when we consider a remotely-located, air-cooled condensing unit. This may be either outdoors or in an unheated area. In the winter, when the outdoor temperature drops to low levels, the condensing temperature will likewise be very low. This, then, may result in a head pressure so low that it prevents proper refrigerant circulation.

Peripheral Devices

Thus, there is a need for a device that will hold the condensed-refrigerant temperature and pressure up to a level sufficiently high to maintain the refrigeration system in proper operation under adverse conditions. The valves used for this purpose are called condenser pressure regulators, Fig. 5-2. They are identical, except for the pressure spring (or high-pressure pilot), to the evaporator and suction pressure valves, and operate similarly. These valves are responsive to compressor discharge pressure, or condenser inlet or outlet pressure, depending on the location of the valve in the condenser circuit.

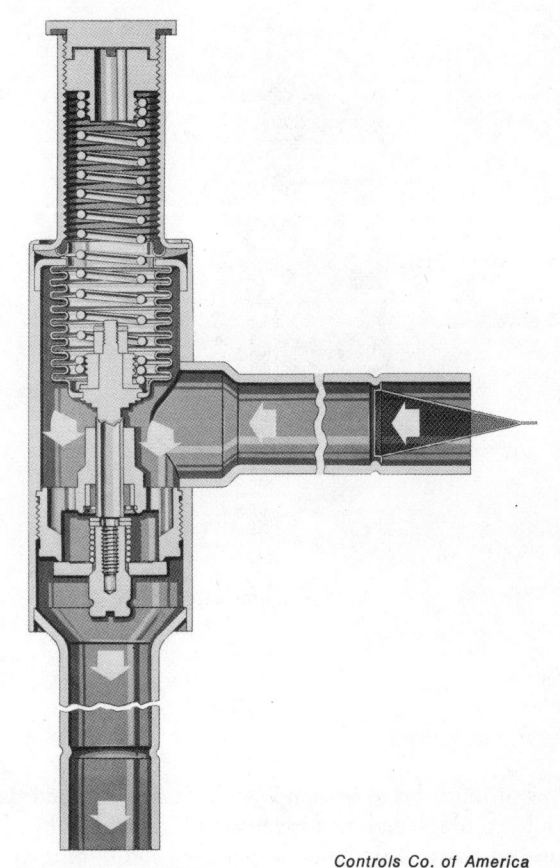

Controls Co. of America
Fig. 5-2. Air-cooled condenser pressure regulator.

Water-Cooled Condenser

Valves designed for use on systems using water for the condensing medium are termed water valves. Their purpose is simple: to control the flow of water through the condenser to maintain constant head pressure.

The action of the water valve, Fig. 5-3 is controlled by the refrigerant pressure within the condenser. Gas from the condenser enters the valve and exerts pressure on the bellows. As the pressure increases, the valve stem connected to the bellows is forced down, thus opening the valve and allowing more water to flow. As the pressure on the bellows decreases, the valve spring forces the valve closed, thus cutting off the supply of water. The pressure at which the condenser is maintained can be controlled by adjusting the pressure of the spring.

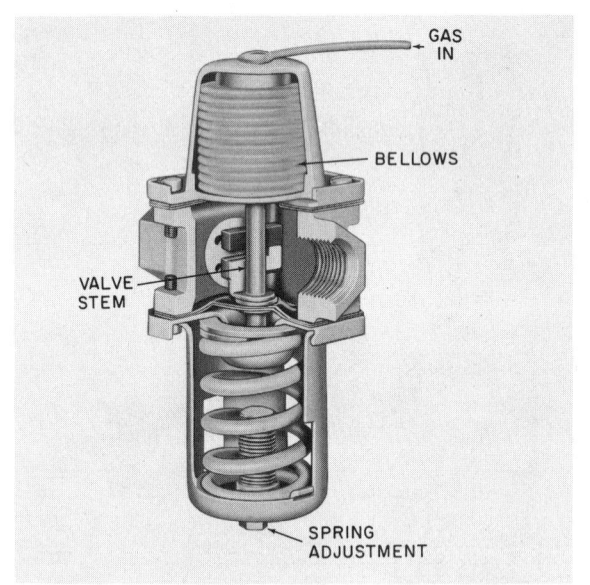

Controls Co. of America
Fig. 5-3. Water valve.

Shut-Off Valves

Manually Operated Valves

The flow of fluids between components in any air conditioning system is contained by piping, and is directed and regulated by the action of valves. In most of the well designed larger air conditioning systems, manually operated shut-off valves are used at strategic locations and greatly assist in system maintenance. These basic valves are used specifically for many applications:

1. Pumping down to replace expansion devices and filter-drier, or cleaning protective strainers
2. Isolating system components for servicing

3. Bypassing automatic controls, solenoids, regulators, and check valves while evacuating, flushing, or drying systems
4. Servicing pressure gauges on control boards
5. Isolating one or more cooling units in multiple-temperature systems
6. Adding safety and protection to equipment and personnel
7. Rapid charging, purging, or equalizing oil levels

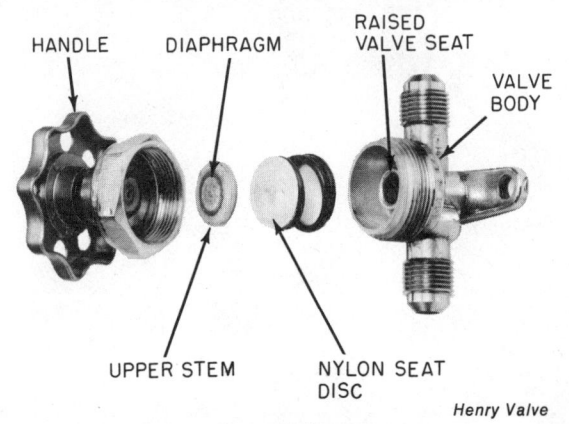

Fig. 5-4. Manual shut-off valve.

Henry Valve

These valves are generally a positive back-seating type with a round easy grip hand wheel, Fig. 5-4. When the hand wheel is moved to the fully-closed position the spring-retained nylon seat disc surface is pressed against the raised valve seat, completely cutting off fluid flow. Pressure from the upper stem working through the non-rotating metal diaphragm keeps the valve closed.

Solenoid Valves

Solenoid valves perform in the same manner as the manually-operated shut-off valves, except that they are electrically actuated. The solenoid coil is a simple form of electromagnet, Fig. 5-5, consisting of a coil of insulated copper wire or other suitable conductor which, when energized by the flow of an electric current, produces a magnetic field that will attract magnetic materials. In this way an armature or plunger can be drawn up into the core of the solenoid. By attaching a stem and pin to this plunger, the valve can be opened or closed by energizing or de-energizing the solenoid coil.

Solenoid valves can generally be divided into two categories: direct-acting and pilot operated. In the direct-acting type, the pull of the solenoid coil opens the valve port directly by lifting the pin out of the valve seat. Since this type of valve depends solely on the power of the solenoid for operation, its port size for a given operating pressure differential is limited by practical limitations of solenoid size.

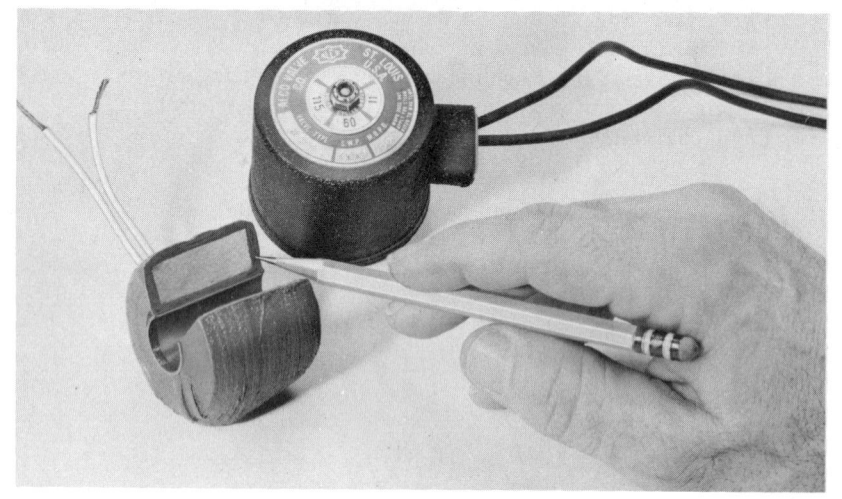

Fig. 5-5. Solenoid valve coil.

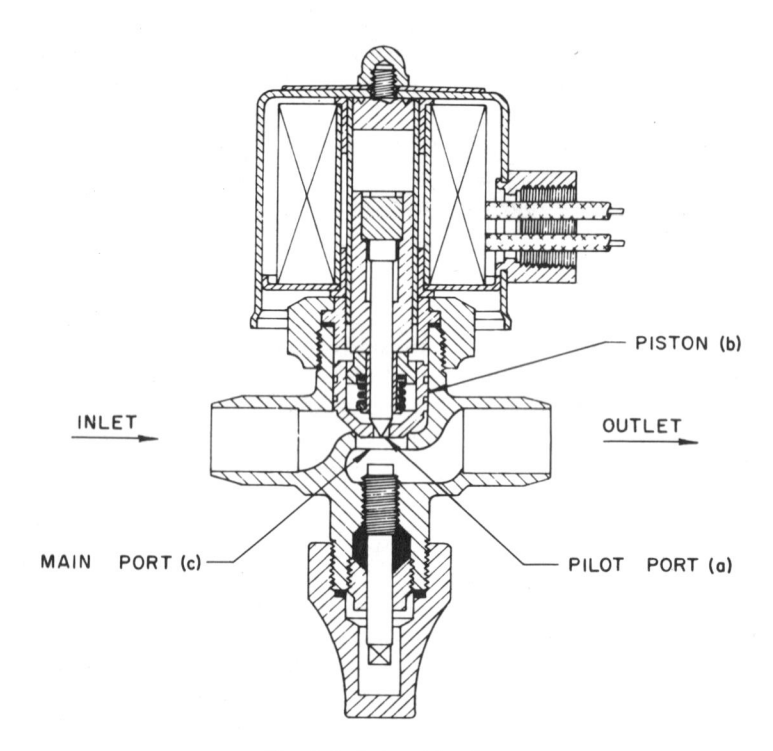

Fig. 5-6. Solenoid valve.
Alco Valve

Larger-size solenoid valves are generally of the pilot-operated type. In this type the solenoid plunger does not open the main port directly, but merely opens the pilot port (a) of Fig. 5-6. Pressure trapped on top of piston (b) is released through the pilot port, thus creating a pressure unbalance across the piston. The pressure underneath is now greater than that above and the piston moves upward, thus opening the main port (c). To close, the plunger drops and closes pilot port (a). The pressures above and below the piston equalize again, and the piston drops to close the main port. The pressure difference across the valve, acting upon the area of the valve seat, holds the piston in a tightly closed position.

Solenoid valves can be used for flood-back prevention, capacity regulation, temperature control, and liquid level control. Selection for a particular control application requires the following information:

1. Fluid to be controlled
2. Capacity needed
3. Maximum operating pressure differential (MOPD)
4. Maximum working pressure
5. Electrical characteristics

All solenoid valves are rated in terms of the maximum operating pressure differential against which the valve will open. For example, with the valve closed and an upstream pressure of 150 psi against a downstream pressure of 50 psi, the pressure differential across the valve will be $150 - 50 = 100$ psi. The MOPD of the solenoid valve to be selected must be equal to or exceed this value.

Reversing Valves

Reversing valves are the most feasible method for changing direction of flow in heat-pump systems (systems that both heat and cool). In the heating cycle of a simple reverse-cycle heat-pump system, the refrigerant flow is such that the indoor coil (normal evaporator) functions as a condenser and provides heat to the indoor air. The outdoor coil (normal condenser) functions as an evaporator, and extracts heat from the outdoor air. Since the outdoor air contains heat down to the lowest temperature at which a refrigerant can function, it can be used as a heat source for the range of winter temperatures normally encountered.

In the cooling cycle, the refrigerant flow is reversed. Since the compressor is a one-way device, a valve is required for changing the direction of flow. This type of valve is called a reversing valve. These valves are generally three- or four-way, two-position valves which are usually operated by pilot solenoid valves. A three-way pilot solenoid valve may be used in the form of a separate valve, or it can be an integral part of the main valve. The pilot valve directs the actuating pressures from the compressor discharge and suction lines to the top of the main valve piston.

Figure 5-7 shows a typical four-way reversing valve installed in a basic heat pump system cooling cycle. This valve may be installed in any position. In this configuration, the discharge gas flows through port (d) to port (1) of the main reversing valve, making the indoor coil act as the evaporator. The suction gas flows from the outdoor coil (in this case the condenser) to the compressor through ports (2) and (s) in the valve. This is the direct opposite of the heating cycle.

The three-way pilot solenoid valve controls the action of the main valve. In the position shown in the cooling cycle, the solenoid coil is energized. Port (a) is closed and port (b) is opened to return. In the reversing valve the slide is down, sealing off nose valve (e) and opening nose valve (c). Thus, nose valve (e) is exposed to suction pressure and nose valve (c) is sealed off from suction pressure by the action of the slide in the pilot valve. Controlled leakage of the high-pressure discharge gas around the reversing valve slide builds up on both ends of the slide, but since there is suction in the area of nose valve (e), the force of the discharge gas is unbalanced, holding the slide in the down position.

To change to the heating cycle, the solenoid in the pilot valve is deenergized, as shown in the heating cycle of Fig. 5-7, allowing its slide to drop. The action is smooth and instantaneous, and the unbalanced pressure holds the slide in the up position until the solenoid is again energized.

Changing cycles from cooling to heating, or vice versa, involves no more than throwing a switch to energize or deenergize the solenoid in the pilot valve. The pilot valve thus is a convenient means for controlling relatively high pressure with a small force.

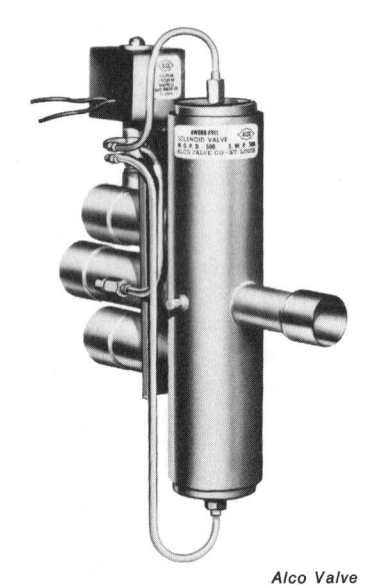

Alco Valve

Fig. 5-7. Four-way reversing valve in heating and cooling cycles.

COOLING CYCLE

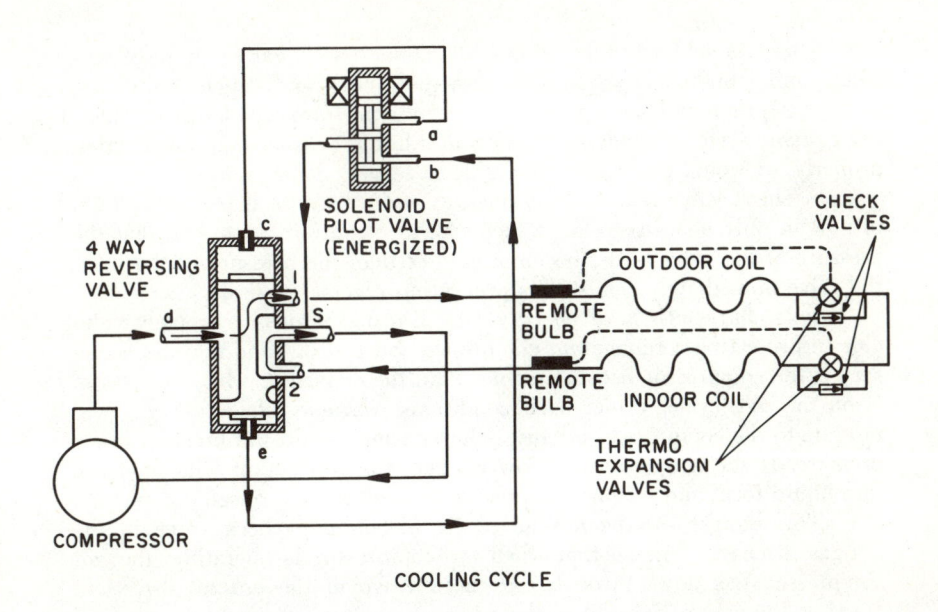

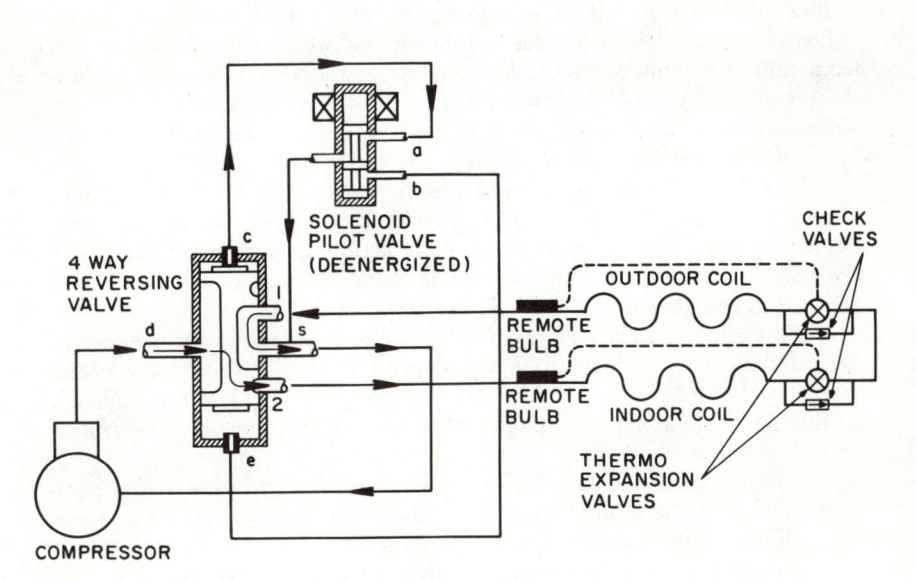

HEATING CYCLE

Check Valves

There is seldom much discussion about check valves in most air-conditioning manuals, perhaps because their need and dependability are taken for granted. However, without a reliable, properly-designed check valve, many systems involving the flow of a liquid or gas could not function properly, or would be highly vulnerable to serious damage.

A check valve is actually a one-way valve because it permits a fluid to flow in only one direction. If a change in conditions occurs so that the pressure at the valve outlet becomes greater than the pressure at the inlet, the valve immediately closes, thus preventing reversal of flow.

This characteristic of check valves also makes them extremely valuable in protecting a refrigerant compressor. Such protection may be necessary if the compressor becomes cooler than the condenser when the system is on the OFF cycle. Under such conditions, warmer refrigerant gas will migrate to the compressor, because a lower temperature will create a lower pressure. If the temperature is low enough, the refrigerant will condense into liquid form and serious compressor damage can be caused on start-up.

This danger can be minimized by installing a check valve in the hot-gas discharge line so that when the compressor is operating, the hot compressed gas flows through the check valve in the normal direction. If, after the compressor stops, the pressure reverses, the valve closes and prevents refrigerant gas backflow to the compressor.

A check valve designed specifically for such hot-gas line applications is illustrated in Fig. 5-8. It is constructed with a cast bronze body and bolted bonnet, and incorporates a Teflon® seat which can withstand high temperatures without excessive deformation. Exact clearances are built into the piston and cylinder so that the piston is aligned perfectly with the valve body seat. Except for piston rotation, this alignment permits reseating at the same point, thus eliminating furrowing and leakage.

The close clearances between piston and cylinder and exact sizing of the piston bleeder opening also prevent clatter and hammering when the valve opens. This is due to the dashpot action of the trapped gas above the piston. This action is featured only in check valves designed for hot-gas lines; it is not used in the liquid-service type of valve.

Smaller type check valves, Fig. 5-9, employ a spring or magnet to help close the valve, while larger sizes depend on gravity and back pressure for valve closing. The larger sizes also have the piston capped and soldered, so that the dashpot action is increased to prevent snap openings regardless of sudden pressure changes.

The hot-gas type check valve must be installed in a horizontal position, and should be located at some distance from the compressor to avoid the pulsating effect of compressor discharge.

Liquid-line type check valves are generally straight-through flow types, and can be installed in any position.

Peripheral Devices 1-93

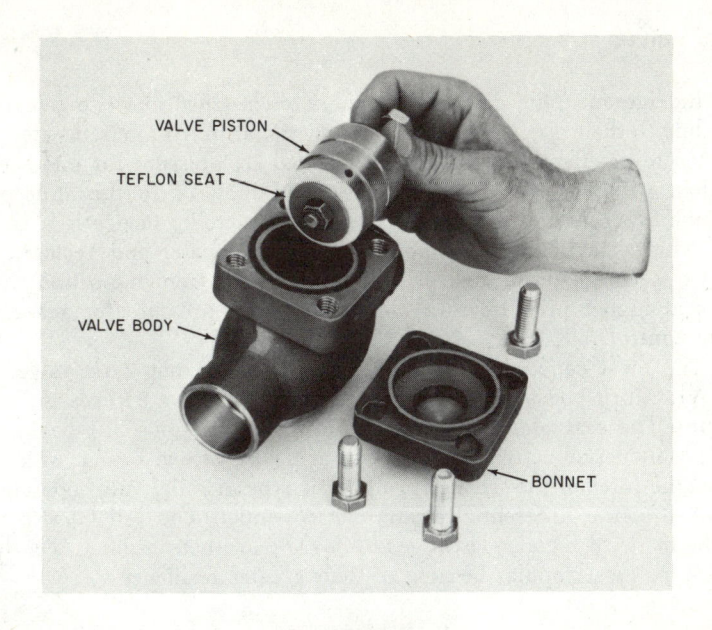

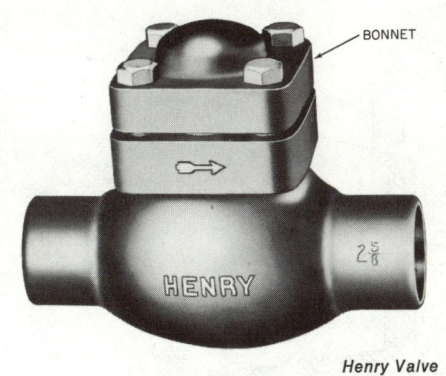

Fig. 5-8. System check valves.

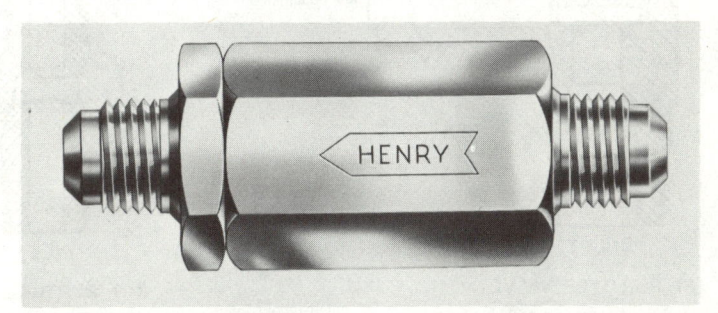

Fig. 5-9. Small-type check valve.

Relief Valves

Refrigerant relief devices are generally classified into two categories, according to their use: either safety or functional. A SAFETY relief device is designed to positively relieve at a designated set pressure for one specific occasion, without prior leakage. This relief may be to the atmosphere or to the low side of the system. A FUNCTIONAL relief device is a control valve which may be called upon to open, modulate, and reclose, with repeated accurate performance. Relief is usually from a portion of the system at higher pressure to a portion at lower pressure for reasons of system control.

The most common safety relief device is the pop type valve, Fig. 5-10(A), which opens abruptly when the inlet pressure exceeds the outlet pressure. The seat of the valve is so designed that once lift begins, the resulting increased active seat area causes the valve seat to pop wide open against the force of the setting spring. This type of relief valve operates on a fixed pressure differential from inlet to outlet. The relief valve seats may be made of metal, plastic, lead alloy, or synthetic rubber. The latter two are the most popular because of their greater resilience.

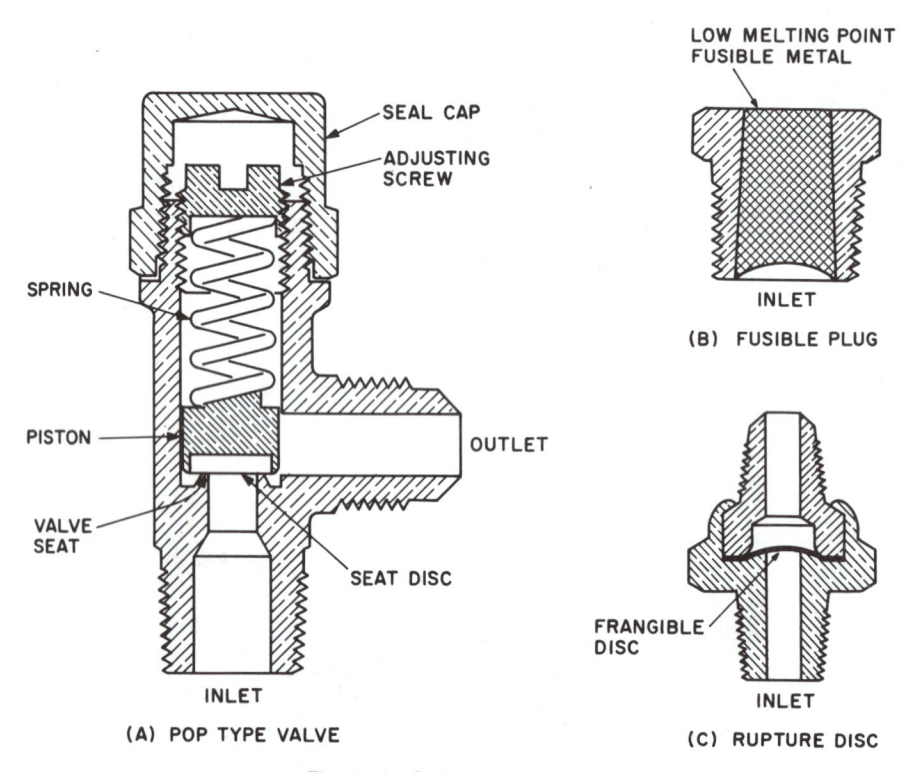

Fig. 5-10. Safety relief devices.

Peripheral Devices 1-95

Other types of safety relief valves are the fusible plug, Fig. 5-10(B), and the rupture disc, Fig. 5-10(C). The fusible plug contains a fusible member that melts at a predetermined temperature corresponding to the safe saturation pressure of the refrigerant, but is limited in application. The rupture disc contains a frangible disc designed to rupture at a predetermined pressure.

Functional relief devices are usually diaphragm type valves, Fig. 5-11, in which the system pressure acts upon a diaphragm which lifts the valve disc from the seat. The other side of the diaphragm is exposed to the adjusting spring and to atmospheric pressure. The ratio of effective diaphragm area to seat area is high, so the outlet pressure has very little effect upon the operating point of the valve.

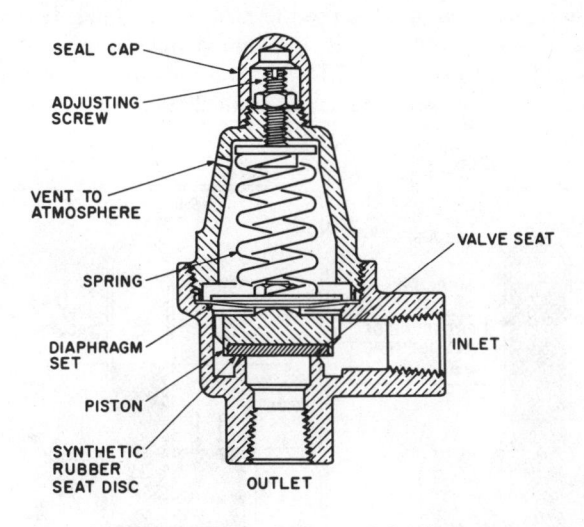

Fig. 5-11. Diaphragm relief valve.

Pressure Gauges

Pressures Below Atmospheric

Barometers are used to measure pressures below atmospheric (below 14.7 psia). Two types of barometers are commonly used: mercury barometers and aneroid barometers.

Mercury Barometer

The mercury barometer indicates pressures in inches of mercury (in. Hg) rather than pounds per square inch gauge (psig).

The simplest form of mercury barometer is the liquid manometer, which consists of a glass tube approximately 33 inches long and closed at one end. It looks like an oversized test tube. The tube is filled with mercury, then quickly inverted and set into a dish or bowl containing additional mercury, with no air allowed to enter the tube. Eventually, the mercury in the tube settles into the bowl, leaving a vacuum chamber between it and the closed end of the tube, Fig. 5-12(A).

At sea level, the mercury in the tube will settle at a height of 29.92 inches. This column is supported only by the pressures of the atmosphere exerted on the surface of the mercury in the bowl. That is, 14.7 psia, or 0 psig. Graduations on the glass tube enable determination as to whether the pressure has dropped (mercury column settles farther down) or gone up.

The barometer can also be used to measure vacuum. If the barometer is in a sealed space, and the air is pumped out of this space until the pressure is only 9.8 psia, Fig. 5-12(B), the mercury column would drop to 19.92 in. Hg. By subtracting 19.92 from 29.92, the amount of vacuum is 10 in. Hg.

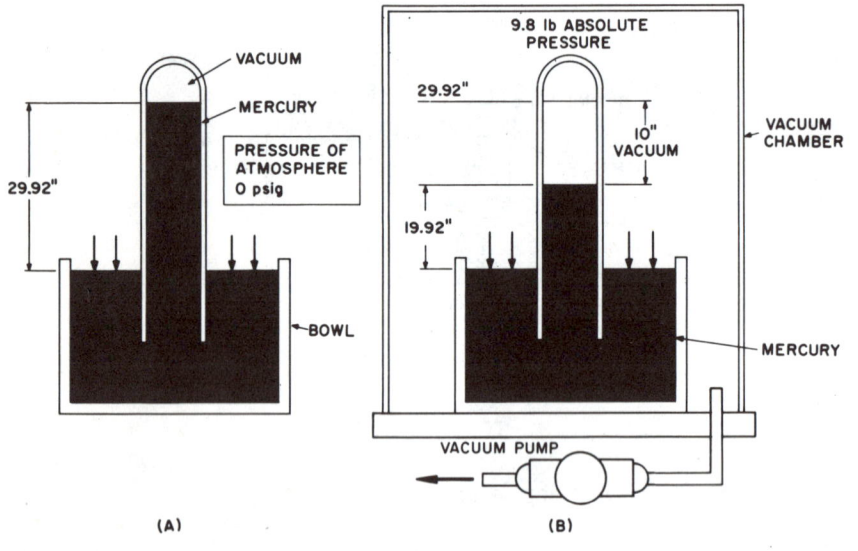

Fig. 5-12. Measurement with a mercury barometer.

In refrigeration work, the barometer in Fig. 5-12 is not convenient for practical use. One disadvantage is its bulkiness. The U-shape of Fig. 5-13 eliminates the need for an immersion tank to hold the excess mercury needed to transfer air pressure to the mercury in the column. The mercury reservoir is part of the tube itself. Some models can be obtained that are calibrated in centimeters and millimeters for use where greater accuracy of measurement is required.

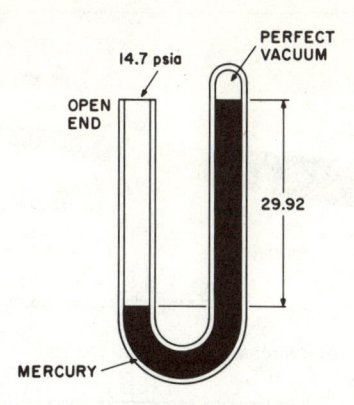

Fig. 5-13. U-shaped manometer.

Another disadvantage of the mercury barometer is that it can be used only in a vertical position. The aneroid barometer, on the other hand, contains no liquids, is compact, portable, and looks like a gauge.

Aneroid Barometer

The aneroid barometer has a sealed metal container which has been highly evacuated, but not quite to a vacuum. Changes in pressure cause the top of the container to respond in either an up or down movement. By means of a series of levers and springs which amplify this motion, a pointer can be made to move in an appropriate direction. Hence, pressure can be read by adding a calibrated scale.

Pressures Above Atmospheric

Three of the more common gauges used in air conditioning work are the diaphragm-operated, the bellows-operated, and the Bourdon-Tube-operated types.

Diaphragm-Operated Gauge

A typical diaphragm-operated type of gauge, Fig. 5-14, consists of a thin piece of metal, the diaphragm, which is very sensitive to pressure changes. Some sort of restrainer, such as a spring, prevents movement of the diaphragm when there is no change in pressure. Any changes in pressure are indicated by a pointer connected either directly to the diaphragm, or by a gear mechanism, or by a gear merchanism which amplifies the distance moved by the diaphragm.

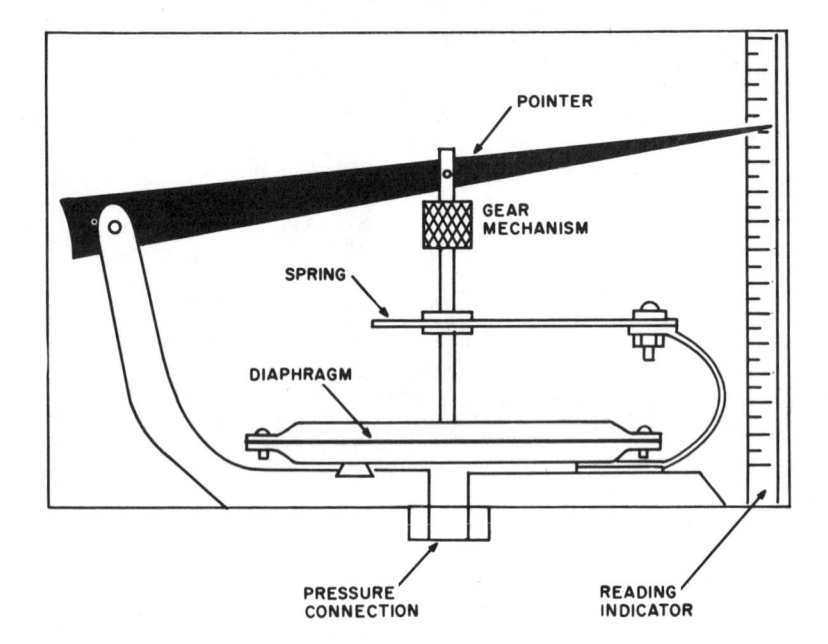

Fig. 5-14. Diaphragm-operated pressure gauge.

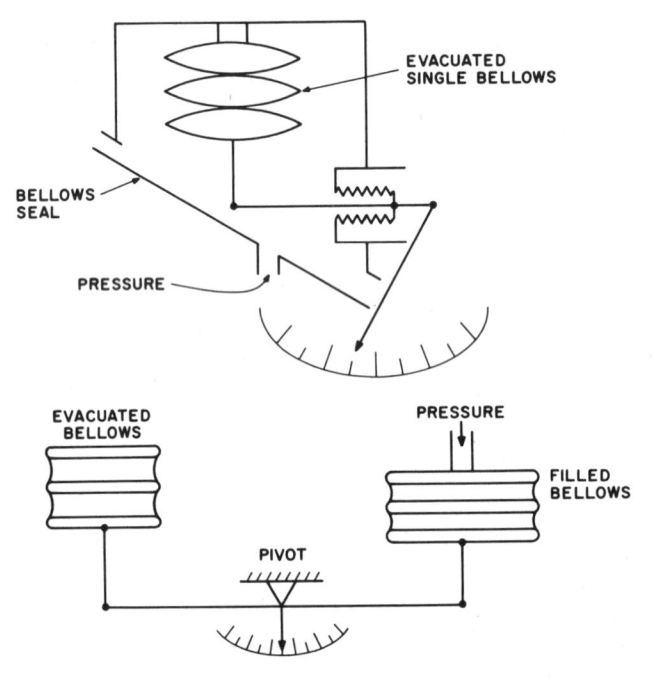

Fig. 5-15. Single- and double-bellows pressure gauges.

Bellows-Operated Gauge

The bellows type gauge can utilize one or two bellows. In the single-bellows type, Fig. 5-15, the bellows is usually evacuated. Incoming pressure causes it to contract. The contraction, in turn, actuates the measuring pointer. This measurement is in terms of ABSOLUTE PRESSURE, as it measures pressure against a vacuum.

In the double-bellows type, one bellows is filled, the other evacuated. Both are balanced on a pivot. The filled bellows also contains the pressure connection. Depending upon an increase or decrease in pressure, the assembly unbalances in one of two directions. The measuring pointer is attached to the pivot, and swings in the same direction of unbalance, again indicating absolute pressure.

Bourdon-Tube-Operated Gauge

The most commonly used type of pressure gauge is the Bourdon-Tube type, Fig. 5-16. All Bourdon-Tube gauges are similar in construction and principle of operation, although they may be used for indicating various amounts of pressure. Rigid tubing, or a combination of rigid and flexible tubing, is used to connect to the inlet (X) at the back of the case. The pressure goes through the tubing and into the Bourdon Tube (M) at (N). This tube is elliptical in cross section, sealed at the outer end, and is made of phosphor bronze or beryllium copper. The outer end is free to move and is connected to a link, lever, and pinion which control the pointer. The end of the Bourdon Tube, which is fastened at (N) to the instrument case is stationary at all times. Because of its construction, the Bourdon Tube acts like a spring and tends to straighten out when internal pressure is applied. This straightening tendency is resisted on the outside surface

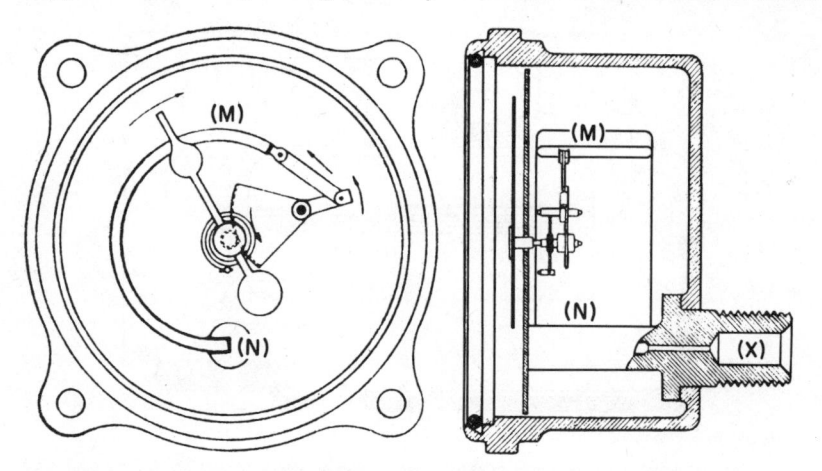

Fig. 5-16. Bourdon tube pressure gauge showing (M) Bourdon tube, and (N) its connection to instrument case.

of the tube by atmospheric pressure, which is admitted to the case by a small vent in the bottom of the case, or by pressure introduced by means of the tubing connection. The gauge indicates the difference between the vent pressure and the pressure inside the tube; hence, it is a DIFFERENTIAL-PRESSURE measuring device.

Compound Pressure Gauge

Another popular type of gauge is the compound pressure gauge, which can measure both pressure and vacuum. Where a standard gauge is used only in a pressure line, the compound gauge can be used in a pressure-suction line. Thus, if the line to which the gauge is attached gives a reading of less than atmospheric pressure, the process occurring at the time is one of evacuation or suction. This gauge reads both in psig and in in. Hg, i.e., pressures above and below atmospheric.

McLeod Gauge

For applications requiring the measurement of high vacuums, the McLeod gauge is used. Hermetically sealed air conditioning and refrigeration systems require a very high degree of moisture and air evacuation. Modern techniques and high-vacuum pumps create extremely high vacuums, within fractions of an inch of a perfect vacuum. This would give only a zero reading on a barometer. Hence, the use of the McLeod gauge, which is calibrated in MICRONS OF MERCURY.

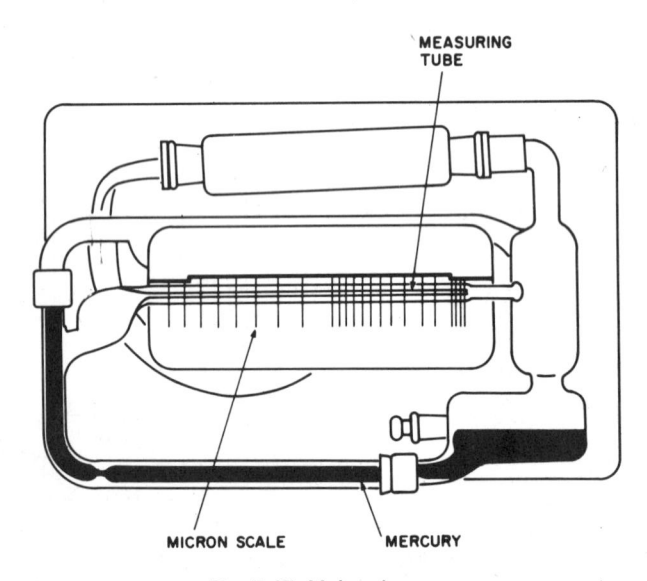

Fig. 5-17. McLeod gauge.

A micron is equivalent to one-millionth (1×10^{-6}) of a meter, or approximately forty millionths (40×10^{-6}) of an inch. Therefore, if a McLeod gauge reads 100 microns, the absolute pressure would be 0.004 in. Hg. This, subtracted from atmospheric pressure would indicate that 29.916 in. Hg vacuum had been obtained (29.92 in. Hg — 0.004 in. Hg = 29.916).

The principle of the McLeod gauge is pressure multiplication; that is, a measured volume of gas is compressed, at a constant temperature, into a volume which is many times smaller than the original. This raises the pressure within that volume, according to Boyle's Law, to within the range of an ordinary mercury barometer.

A modification of the McLeod gauge, Fig. 5-17, uses mercury as the compressing agent. The unit consists of a gas-filled graduated measuring tube, another tube containing mercury, and the pressure connection. When the pressure enters the mercury-filled tube, the mercury, in turn, compresses the gas in the measuring tube. The pressure is then read from the graduation which the mercury has reached in the measuring tube.

Filter-Driers

The primary purpose of a filter-drier is the removal of moisture from the oil-refrigerant mixture flowing in the liquid line of a refrigeration system. The amount of moisture in a system must be kept below an allowable maximum in order to prevent freezing at the expansion device, corrosion of metals, formation of sludge, valve failure, or chemical damage to insulation or other system materials.

Many filter-driers can also filter out refrigerant contaminants without adding undue resistance to the flow. Contaminants in a refrigeration system are those substances which serve no useful function, and may be injurious and interfere with proper operation. They include not only dirt and moisture which may remain after system manufacture and installation, but also acid, sludge, and other products of the chemical reactions which take place while the system is operating, see Fig. 5-18.

Modern desiccants in filter-driers easily make an acceptable dry system possible. A DESICCANT is a substance capable of removing the moisture from a gas, liquid, or solid. Although this is its basic characteristic, there are several important secondary properties a dessicant must have:

1. It must be able to reduce the moisture content to a level of approximately 5 parts per million (ppm).
2. It must be able to absorb the moisture found in the installation, and still retain its efficiency.
3. It must act rapidly; that is, it must reduce the moisture sufficiently in one passage through the drier unit, so that the refrigerant leaving the unit will not have enough moisture left in it to freeze up the expansion valve or capillary tube of a system operating at low temperatures.

Basic Air Conditioning

Fig. 5-18. Contaminants trapped within filter-dryer.

4. Its efficiency, capacity, and speed should not seriously be affected by increases in temperature up to approximately 125°F, or by the presence of oil or other substances that might be added to a refrigerant system.
5. It must retain all of its characteristics important to proper functioning during both storage and operation.
6. It must not dissolve in any of the liquids commonly found in a refrigeration system: it must remain in its original solid condition and at no time be reduced in size or consistency.
7. It must be of such form as to properly carry out its functions. If granular or pelleted, the granules must be of such a size as to give proper contact with the refrigerant, while still allowing adequate flow. If it is formed in a solid block core, it must be sufficiently porous to allow ready flow of the refrigerant through it with little restriction or pressure drop.
8. It must not be hazardous in either storage or in use: it must not be poisonous or irritating, nor should it react chemically with moisture to produce products that are dangerous or corrosive.

Desiccants can be molded into a special form to fit a drier container. Such containers can either be sealed or replaceable-core types. Most molded desiccants are formed into a porous block, Fig. 5-19. They are combinations of molecular sieve drying agents and either silica gel or activated alumina. Silica gel is manufactured as a jelly-like form of silica acid or silicon dioxide. It is subjected to a partal drying process when manufactured. As the gel begins to lose water, it develops a tremendous surface area. These surfaces are made up of very fine capillary tubes. Within the walls of these formations is a minute quantity of water. This water cannot be removed from the gel without ruining the desiccant.

Activated alumina (alumina hydroxide) has a similar structure, and

Peripheral Devices 1-103

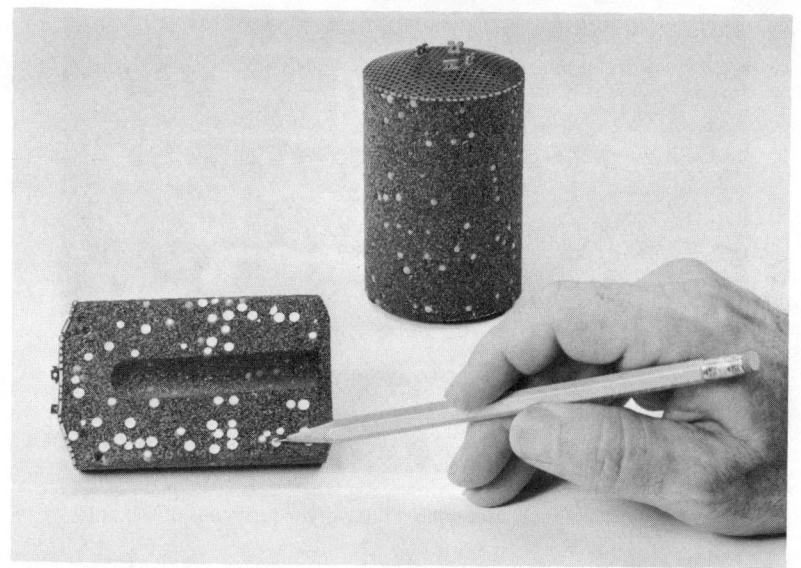

Fig. 5-19. Porous block showing molecular sieve drying agents (molded desiccant). *Alco Valve*

also contains a definite quantity of water which is essential to the desiccant structure. These gel-type desiccants pick up and hold moisture by the physical process of absorption, in which the molecules of water are stuck on the inner walls of the capillaries in almost the same manner as a postage stamp is stuck to the surface of an envelope. Since the surface area of the many capillaries is tremendous, and the size of the water molecule small, a relatively large amount of water may be picked up and held within the body of the gel.

Both silica gel and activated alumina are most effective with liquid refrigerants in the low temperature range. When these drying agents are properly proportioned with the molecular sieve materials, driers with a wide temperature range and large drying capacity are produced.

When installed in sealed containers, Fig. 5-20, desiccants are subjected to carefully controlled conditions of temperature and high vacuum. They are completely activated within the container to insure maximum efficiency.

Replaceable-core driers, Fig. 5-21, are used to cut handling time during replacement to a minimum. In these units, two replaceable cores are fitted into place and held there by either spring action or bolt action. The cores are generally fitted over a stainless steel strainer that acts as an assembly spool to hold the blocks in place. Except for their shape, the replaceable type blocks are basically the same composition as the sealed type.

Driers using granular type desiccants are generally all sealed units, Fig. 5-22. Granules of activated alumina are tightly packed within a sealed

1-104   *Basic Air Conditioning*

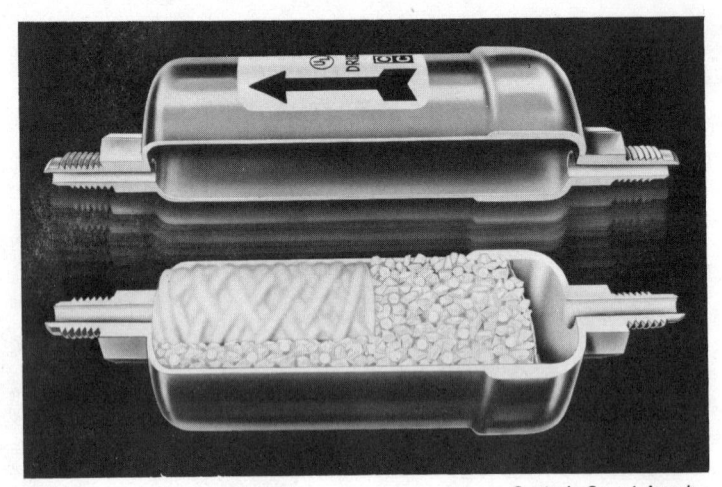

Controls Co. of America
Fig. 5-20. Cross section of a typical filter-dryer (granular desiccant).

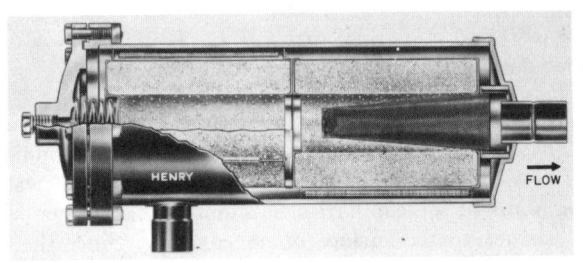

Henry Valve
Fig. 5-21. Replaceable core filter-dryer.

container which permits intimate contact with the refrigerant for quicker absorption of water and acids.

Manufacturers rate their driers for moisture capacity in accordance with Air Conditioning and Refrigeration Institute Standard 710-58. This standard makes it easy to pick the proper size drier for the required function.

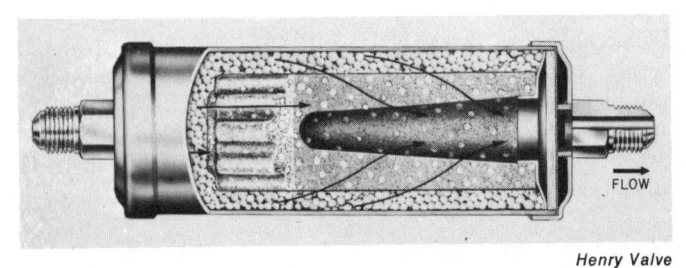

Henry Valve
Fig. 5-22. Seal dried with fluted core and granular **desiccant**.

6
Motors and Controls

Although air conditioning electric motors and their controls belong in the chapter on peripheral devices, the subject is so large and important that it warrants this separate discussion.

Compressor and Fan Motors

Motor Types

Electric motors come in many different types, Table 6-1, for various services. They are classified as either fractional horsepower (hp) or integral hp. Fractional hp motors have ratings of less than 1 hp at 1700 to 1800 rpm and are available in a wide variety of sizes. In refrigeration and air conditioning applications, mostly single-phase fractional hp motors are used. Integral hp motors are largely single-phase up through 6 hp. Larger motors are polyphase or use direct current.

Before discussing the various motors used in air conditioning applications, it would be wise to review their basic principles of construction and operation.

Basic Operation

Operation of electric motors is based on the principles of electricity and magnetism. The latter principles can be demonstrated by suspending a permanent magnet in such a way that it is free to turn, Fig. 6-1. If we bring another permanent magnet close to the pivoted one, the pivoted magnet starts to rotate, because the poles of two permanent magnets will either attract or repel. Like poles repel each other, while unlike poles attract each other. This principle of attraction and repulsion is converted to rotary motion in an electric motor.

In an electric motor, electromagnets supply the magnetic field necessary for operation. The core of an electromagnet is magnetized only when current flows through it. The polarity of an electromagnet can be reversed by reversing the direction of current flow through its coil.

A typical induction motor consists of a stationary section called a STATOR and a rotating section called a ROTOR. The stator is made up of thin steel laminations. When these laminations are wound with insulated

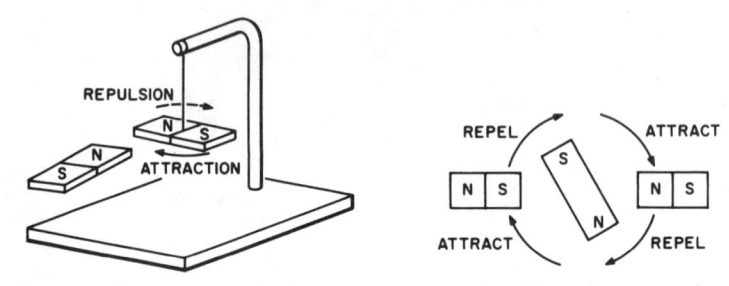

Fig. 6-1. Magnetic attraction and repulsion.

wire, the result is an electromagnet. If a 60-Hz alternating current is applied to the electromagnet, the stator, or field coil as it is more commonly known, becomes an electromagnet whose magnetic poles reverse their polarity 120 times per second; that is, every time the current direction alternates, see Fig. 6-2(A).

If a magnet is placed in the center of the magnetic field and given a spin, and the 60-Hz ac is applied to the stator coils, the magnet, or rotor, will continue to spin, Fig. 6-2(B), due to the attraction and repulsion of the alternating polarities of the stator's poles.

Table 6-1. Motor Types

Type	Horsepower Range	Type Power Supply	Suitable for Hermetic Application
Fractional Horsepower Sizes:			
Split-phase	1/20 – 1/3	Single-phase	Yes
Capacitor-start	1/8 – 1	Single-phase	Yes
Repulsion-start*	1/8 – 1	Single-phase	No
Permanent-split-capacitor	1/20 – 1	Single-phase	Yes
Shaded pole	milli hp	Single-phase	No
Squirrel-cage-induction	1/6 – 1	Polyphase	Yes
Direct-current	1/20 – 1	D-C	No
Integral Horsepower Sizes:			
Capacitor-start/ Capacitor-run	1 – 6	Single-phase	Yes
Repulsion-start*	1 – 5	Single-phase	No
Squirrel cage-Induction (normal torque)	1 – up	Polyphase	Yes
Slip-ring	1 – up	Polyphase	No
Direct-current	1 – up	D-C	No
Permanent-split-capacitor	1 – 6	Single-phase	Yes

Seldom used in air conditioning.

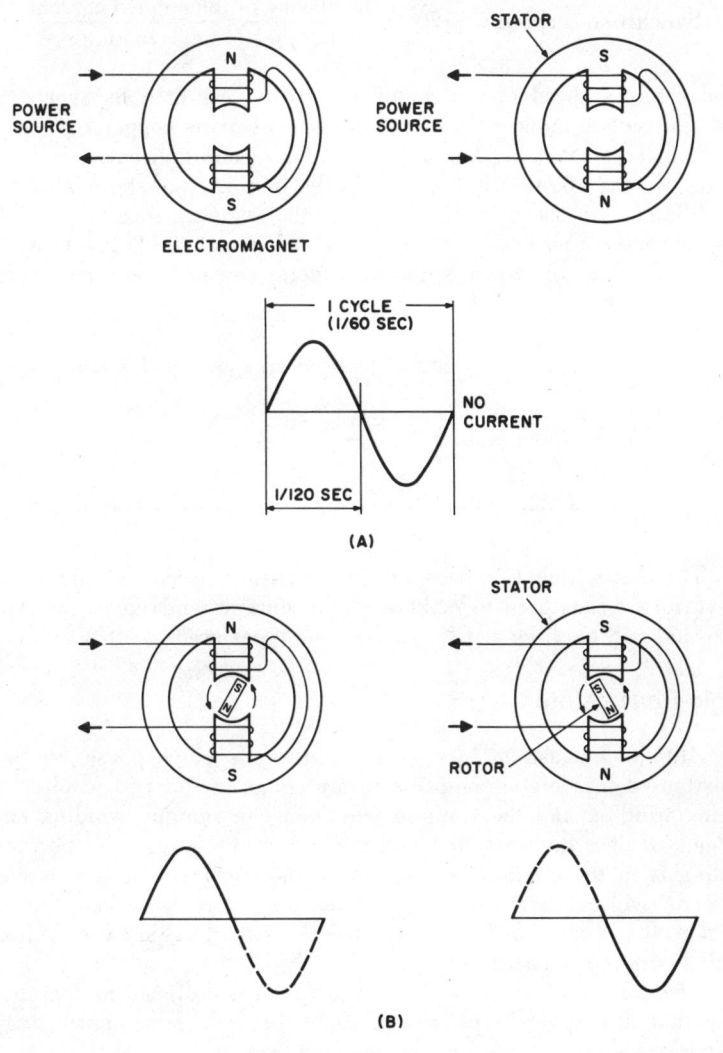

Fig. 6-2. The two-pole stator: (A) As current direction alternates, magnetic poles of the stator change. (B) Once started, permanent magnet spins in magnetic field.

The most common type of rotor is known as the squirrel-cage type, Fig. 6-3. This rotor is composed of slotted sections of thin steel similar to that of the stator. The slots of the rotor contain bars of bare copper, brass, or aluminum shorted at each end with shorting rings.

The rotor will continue to spin at a theoretical speed known as the SYNCHRONOUS SPEED. The synchronous speed of a motor is determined by the use of the following formula:

$$\text{Synchronous speed} = 120 \times \frac{\text{frequency of the applied current}}{\text{number of poles in motor}}$$

Actual running speed of the motor is slightly less than its synchronous speed. Current is induced in the rotor only when its copper bars cut the lines of force created by the stator field. This condition exists only when the rate of rotor rotation is slower than the frequency of stator alterations. This difference in speed is called SLIP. Actual running speed is, therefore, the synchronous speed minus the slip, which is about 4–5% below synchronous speed. Thus, the actual speed of a motor can be determined with the following formula:

$$\text{Actual speed} = \text{synchronous speed} - 4\% \text{ slip}$$

$$(\text{For a 2-pole motor}) = \frac{120 \times 60}{2} - 144 = 3456 \text{ rpm}$$

$$(\text{For a 4-pole motor}) = \frac{120 \times 60}{4} - 72 = 1728 \text{ rpm}$$

The motor that has been under discussion needed a spin to start. The various means used to make a motor start depends upon the type of motor and will be discussed for each of the types used.

Single-Phase Motors

All single-phase motors (motors using a single-phase a-c power source) used in a motor-compressors are comprised of two windings: the starting winding and the running winding. The running winding carries the load, or does the work during the operating cycle, as only the running winding is in the circuit. As long as the electrical circuit is not opened, the motor will continue to operate. When the circuit is broken, the motor will stop. It will not, however, start when the circuit again closes unless the starting winding is energized.

There are four types of single-phase motors designed to operate in a refrigerant atmosphere: split-phase, capacitor-start/induction-run, capacitor-start/capacitor-run, and permanent-split capacitor.

Split-Phase Motor

The split-phase (low-starting-torque type) motor can be used on any installation that has a capillary tube as the metering device. As this type of motor does not have sufficient starting torque to start against a high head (condenser) pressure it can only be used on an installation in which the high- and low-side pressures balance before the start of the running cycle.

In split-phase motors, when the start winding is energized out of phase with the main winding, it produces sufficient magnetic pull to cause

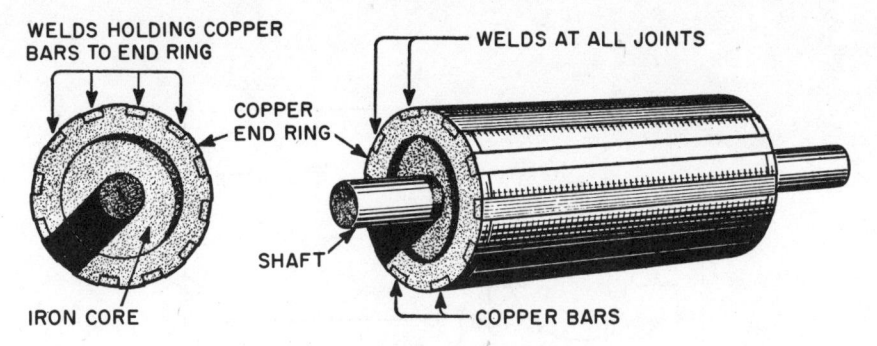

Fig. 6-3. Squirrel-cage rotor.

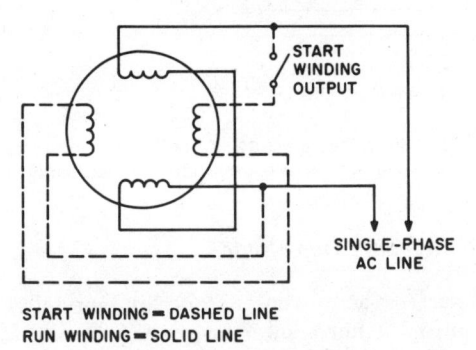

Fig. 6-4. Split-phase motor.

the motor to rotate. This is accomplished by connecting both windings as shown in Fig. 6-4. If this connection were left in the circuit, the start winding would be damaged by excessive heat. Therefore, the start winding is removed from the circuit at approximately 85% of rated speed by a cutout device in the line. This device can be centrifugally operated, or be some sort of relay whose operation depends upon motor speed.

The starting torque of a split-phase motor is low, the starting current is high and the efficiency is relatively low, hence, its restriction to capillary-tube systems in the lower fractional hp range.

Capacitor-Start/Induction-Run Motor

The capacitor-start/induction-run type motor is used where expansion valves are employed as this type of motor can be started even with a fairly heavy load. In order to provide higher starting torque a starting capacitor is provided, as shown in Fig. 6-5. The starting capacitor is removed from the circuit when the motor reaches approximately 85% of rated speed.

Capacitor-start/induction-run motors are used where higher starting torque is required and power factor is not important. Its range is usually between 1/6 to 3/4 hp.

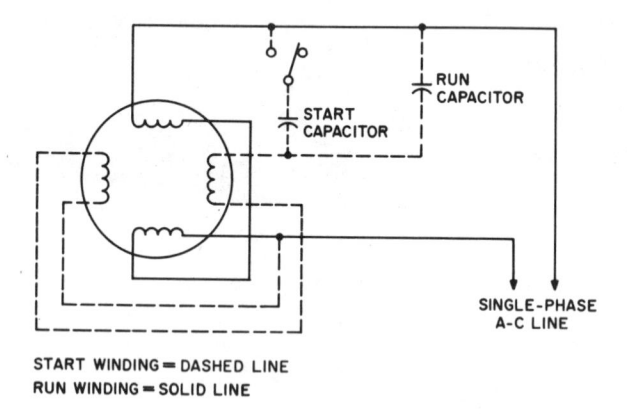

Fig. 6-5. Simplified capacitor-start/induction-run and capacitor-start/capacitor-run motors. The run capacitor is used only in capacitor-run motors.

Capacitor-Start/Capacitor-Run Motor

Capacitor-start/capacitor-run motors are normally used in motor-compressors requiring 1 horsepower or greater. However, it is sometimes used in 3/4 hp and smaller sizes to reduce running current and increase power factor. This type of motor not only has the advantage of higher starting torque, but also will operate under a heavier running load than will the induction-run motor. By connecting the running capacitor in parallel with the starting capacitor, as shown by the dotted lines of Fig. 6-5, the motor is strengthened, because the start winding remains connected in phase with the run winding after the start capacitor is disconnected. Thus, the start winding carries part of the running load. The running capacitor improves the power factor, increases the efficiency, and reduces motor current, thus decreasing the temperature of the motor.

Permanent-Split Capacitor Motor

For some applications not requiring high starting torque, a simple motor with only a running capacitor connected as shown in Fig. 6-6, is desirable. This eliminates the cost and possible service problems of starting capacitors and winding cutout devices, but does not present definite limitations because of its relatively low starting torque. This type of motor is used largely in room air conditioning units and small commercial applications.

A permanent-split capacitor motor, Fig. 6-7, has two windings: the main (running) winding and the starting winding. The running capacitor is connected in series with the starting winding. Both, in turn, are connected in parallel with the running winding, and the capacitor is kept in the circuit during both the start and run phases of the motor. Thus, no start-winding cutout device is necessary.

The capacitor provides the phase differential between the windings, creating the torque required to start the motor. The start winding is of a relatively high resistance as compared to the run winding. Therefore, when the motor approaches operating speed, the current will flow through the lower-resistance run winding more readily. At the same time, the capacitor helps to limit the motor running current because it impedes the current to the start winding, and also helps to increase the power factor of the motor.

As compared to a capacitor-start motor, a permanent-split capacitor motor has a lower starting torque, but has a higher power factor and a relatively low running current.

START WINDING = DASHED LINE
RUN WINDING = SOLID LINE

Fig. 6-6. Permanent-split capacitor motor.

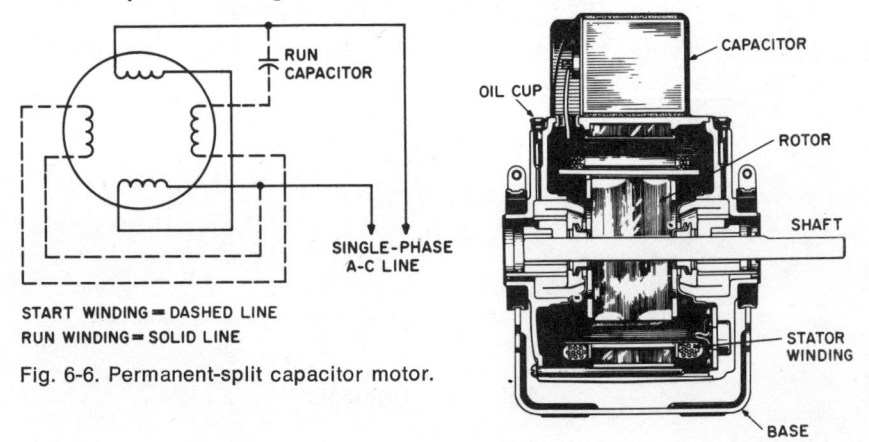

Fig. 6-7. Pictorial view of a permanent-split capacitor motor.
Redmond Motors

Polyphase Motors

Two-Phase Motors

Two-phase motors are used in very few air conditioning applications. They consist basically of two parallel-connected windings. These have high starting torque, and do not require either capacitors or starting relays. They require some sort of circuit make/break device, to break all lines when the control circuit is deenergized.

Three-Phase Motors

Three-phase motors are used in larger compressors. Because of their simplicity, they are used extensively commercially and in central air conditioning systems. Most of these motors are in the 2 hp to 6 hp

1-112   *Basic Air Conditioning*

range. They have sufficient starting torque, high efficiency and power factor, and do not require capacitors or starting relays. They require some sort of make/break start device, depending on the type of protection employed.

Most three-phase motors used by the compressor manufacturers are delta-connected, as shown in Fig. 6-8, or wye-connected. Some motors of 7-1/2 hp and larger have two parallel three-phase windings suitable for winding start, which can be connected in series for operation on higher voltage.

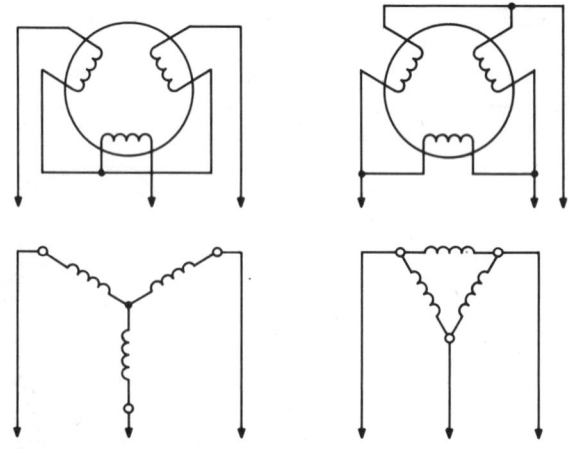

Fig. 6-8. Two 3-phase motors with equivalent wye and delta connections.

Automatic Controls

Purpose

Automatic controls are used for one or more of the following purposes:

1. To control temperature, pressure, or humidity
2. To prevent operation when operation would be hazardous
3. To insure economical operation by providing even operating cycles and preventing superfluous or excessive operation
4. To eliminate human error. To control the various functions and components of a system in a manner impossible to accomplish manually

There are over a thousand different automatic control devices in use, many of which are quite complex. It is impractical to discuss all of them. However, there are a few fundamental controls that are basic to automatically-controlled systems. These devices and their operation are discussed on the following pages.

Capacitors

As mentioned earlier, a capacitor is an electrical device which temporarily stores electricity up to the peak of the voltage cycle, then discharges as the voltage drops from the peak. This storage-discharge function provides the additional power and torque required for starting and running the air conditioning motors.

A capacitor has three essential parts: two metal plates and a dielectric which separates and insulates the two plates from each other.

Three basic factors influence the storage capacity: plate area, dielectric thickness (or volume), and dielectric material. The greater the plate area, and the smaller the distance between plates, the greater the storage capacity. The greater the amount of the dielectric, the greater the capacity.

Capacitors are classified according to dielectric material. Electrolytics are those which use a wet (oil) or dry (tantalum) electrolyte as the dielectric. Paper types are those which use paper as the dielectric. There are others, also, such as ceramic and mica, but these have very small capacities and working voltages as compared to those required for motor operation. These types are used mostly in electronic work.

Start capacitors are usually electrolytics; run capacitors are special paper types. As the names imply, the start capacitor is used to help start the motor; the run capacitor, on the other hand, is used to both reduce the required running current and improve the motor power factor.

Start Capacitor

The start capacitor, Fig. 6-9, provides the phase differential necessary for starting a compressor motor. It temporarily allows a high current, at a phase different from that in the run winding of the motor, to bypass the current running through the run winding, thus creating the necessary starting torque. When a run capacitor is used in conjunction with a start capacitor, the start capacitor is connected in parallel with the run capacitor, thus adding capacity to the circuit. This arrangement also permits increased starting current to flow, and decreases the amount of time that this high starting current is required. A start capacitor is in the circuit from start-up until the compressor motor approaches normal operating speed. Total in-circuit time is only a few seconds. The start capacitor is placed in series with the start winding of the motor, and causes the current in the start winding to lead the current in the run winding (i.e., 90° degrees ahead of the run current).

During the OFF cycle, the voltage that has built up across the start capacitor might discharge through the cutout device contacts. Should this occur the contacts may fuse together, and cause the start capacitor to break down internally because of excessive short-circuit current. To prevent the start capacitor's high voltage from discharging through the cutout contacts, a bleeder resistor (which slows down the discharge) is placed across the start capacitor terminals.

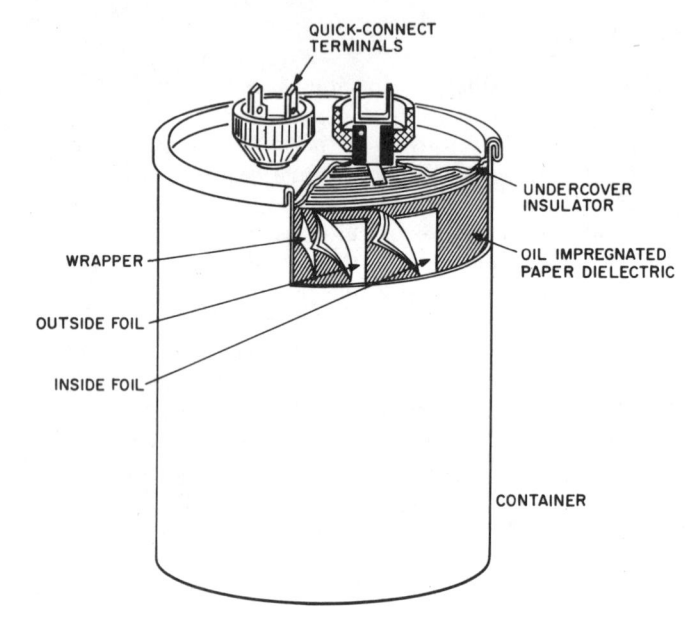

Fig. 6-9. Typical start capacitor.

Run Capacitor

A run capacitor, Fig. 6-10, functions similarly to the start capacitor, but has lower capacity and higher voltage rating so that it may be left in the circuit continuously. It increases the motor power factor, thus decreasing the current requirements.

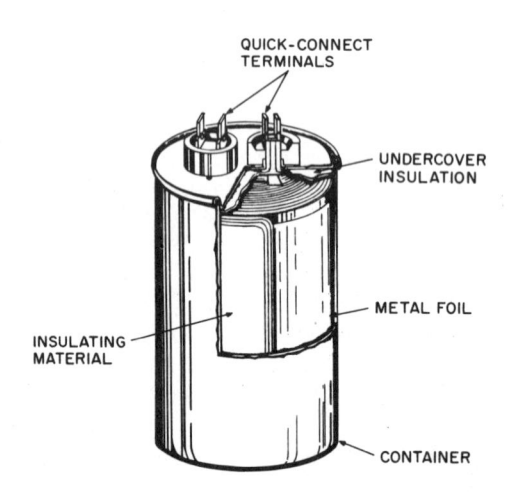

Fig. 6-10. Run capacitor.
G.E.

Motors and Controls 1-115

The run capacitor is of the oil-paper type. It consists of sheets of metal foil that are insulated from each other by layers of oil impregnated paper, and inserted in an oil-filled container. The run capacitor is also connected in series with the start winding, and the combination placed in parallel with the run winding. The result is a reduction in full load current during continuous duty, because of the added start circuit. The oil helps to prevent the capacitor from overheating under normal operation. However, voltage irregularities, continued overloads, and the result of overheating will make this capacitor less effective. This, then, results in full-load current increase.

Electric Relays

Electric relays are used to start and stop motors driving refrigeration compressors, fans, pumps, or other equipment for which the electrical load is too large to be handled directly by a manual control.

Solenoid-Operated Relay

The basic type of electric relay is the solenoid-operated switch shown in Fig. 6-11. When the solenoid coil of the relay is energized, the electromagnetic attraction produced by the coil pulls in an armature which opens or closes the contacts of a second, but independent circuit. Circuit connections are made through the contact screws. This device may be used to control a high voltage or heavy current by means of a small current through the solenoid. It may also be used to interlock several electrical circuits, but would require a greater number of blades to do so.

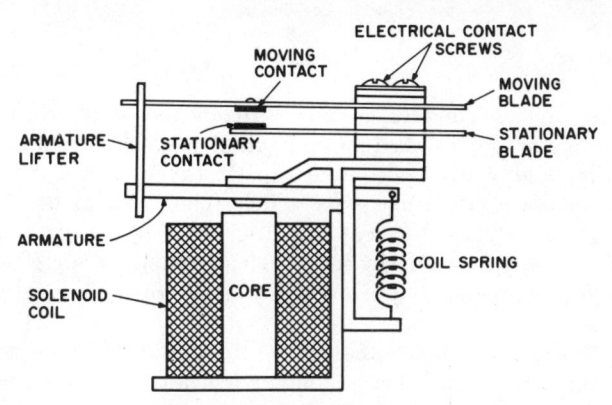

Fig. 6-11. Solenoid-operated relay.

Thermal-Actuated Relay

A variation of the basic electric relay is a relay that is actuated by a thermal rod. These thermal rod actuators, Fig. 6-12, are made of a metal with a high expansion coefficient, and enclosed in a suitable insulator around which resistance heating wire is wound. Current through this wire causes the wire to heat up. This, in turn, causes the thermal rod to expand. The expanding rod pushes a lever against a power switch which controls the voltage applied to the circuit being controlled. The time that it takes for the resistance wire to heat up depends on the size of the resistance wire. Furthermore, the time that it takes for the thermal rod to expand depends on its thermal mass. Therefore, the design is capable of a wide range of time delays and switch actuation arrangements. Its usual use, in air conditioning, is in temperature control.

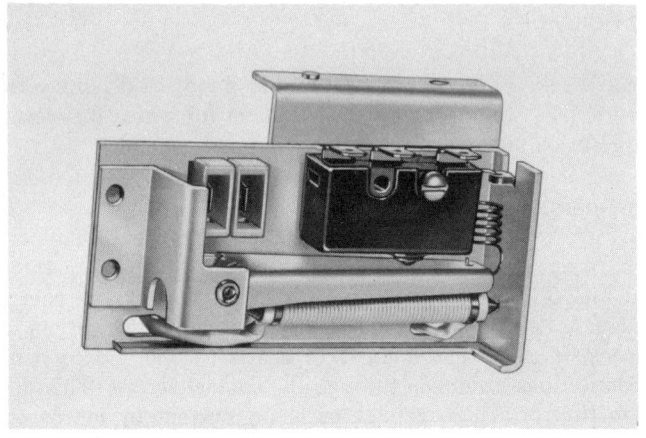

Controls Co. of America
Fig. 6-12. Thermal actuated relay.

Starting Relays

Starting relays conduct current through the start winding of a single-phase motor during start-up and initial run. Motor starting relays are normally used with single-phase capacitor-type motors where it is impractical to use a centrifugal-type switch (connected to the rotor and opened at preset rotational speed) for disconnecting the start capacitor. Such applications as hermetically-sealed refrigeration and air conditioning motors (as in compressors requiring high starting torques) usually require relays of this type.

The starting relay directs the current through the motor's start winding via the start capacior. When the motor reaches a predetermined speed (usually, at approximately 85% of running speed), the contact points of the relay open, dropping the start capacitor out of the circuit.

There are two general types of motor starting relays, the current relay and the potential or voltage relay. The use of either is determined by the motors they are to be used with.

Current Relay

The current relay shown in Fig. 6-13 consists basically of an electromagnetic coil, an armature on a guide pin placed within the coil, and a double set of switch contacts, all within a housing. Figure 6-14 shows a typical air conditioner motor circuit using a current relay. As shown, the relay coil is connected in series with the run winding of the compressor motor, while the normally-open switch contacts are series connected with the start capacitor.

Since the relay coil is connected in series with the run winding, run-winding current will flow through the relay coil during both start and run periods. When the initial voltage is impressed on the motor, the heavy momentary inrush current (starting current is usually 4 to 5 times the running current) in the run winding flows through the relay coil, energizing the coil. This pulls up the armature (solenoid action), which, in turn, causes the normally-open switch contacts to close, thus completing the start capacitor circuit.

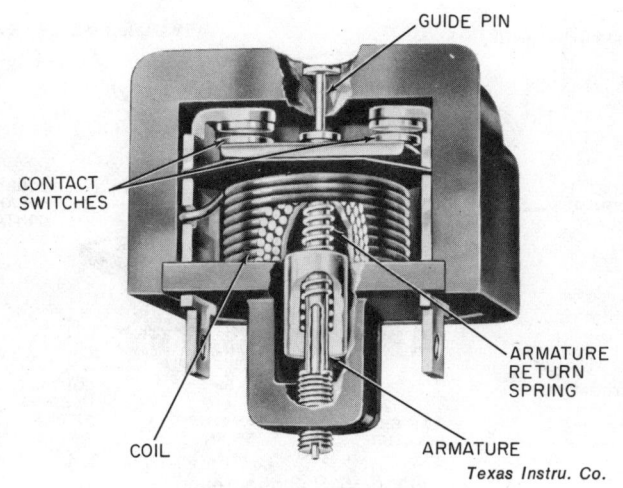

Fig. 6-13. Current relay.

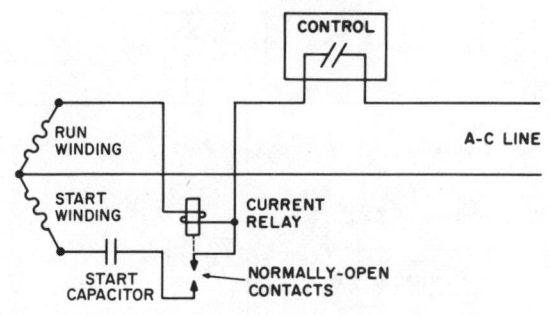

Fig. 6-14. Hermetic circuit with current relay.

As the motor approaches its operating speed, the current in the run winding decreases, lessening the current in the relay coil. At approximately 85% of the motor's operating speed, the current in the relay coil has decreased to a point where the compression of the armature return spring is too great to be held in check. The armature then drops back to its original lower position which, in turn, causes the switch contacts to open and drop the starting capacitor out of the circuit.

Potential or Voltage Relay

The potential or voltage relay also keeps the start capacitor in the circuit momentarily, but does so in a different way.

The potential relay shown in Fig. 6-15 consists basically of an electromagnetic coil, a leaf-shaped armature, and a double set of switch contacts, all within a housing. Figure 6-16 shows a typical air conditioner motor circuit using a potential relay. As shown, the relay coil is connected in parallel with the motor start winding, while the normally-closed switch contacts are connected in series with the start capacitor.

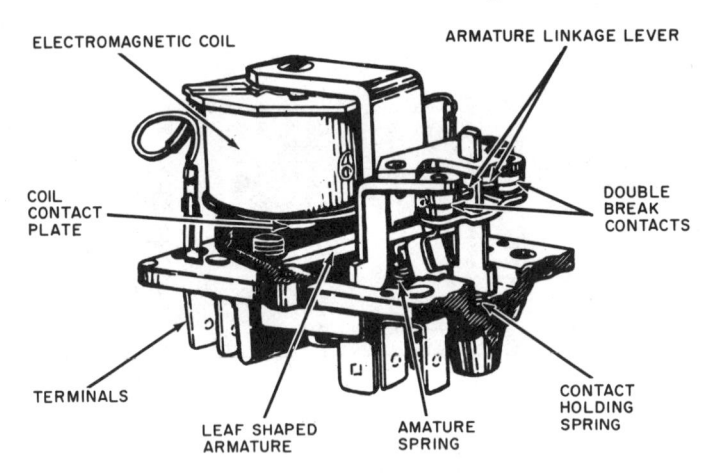

Fig. 6-15. Potential relay.

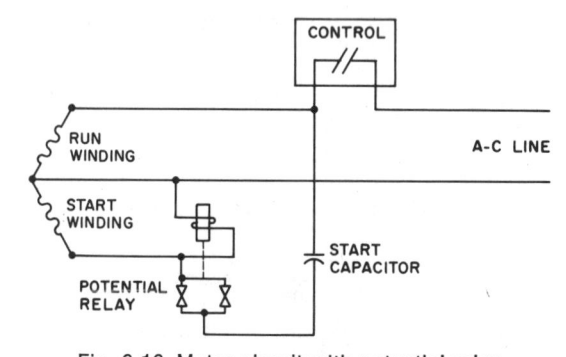

Fig. 6-16. Motor circuit with potential relay.

Motors and Controls 1-119

The contact points on this relay are held normally closed by a spring. As current starts to flow through the run and start windings, it is immediately directly through the closed contacts and through the start capacitor. The voltage across the relay coil (which is connected in parallel with the compressor motor start winding) increases with the increase of speed of the motor. As the motor approaches operating speed, the voltage induced in the coil causes solenoid action, and the armature is pulled up. The armature linkage arrangement, in turn, forces open the contact points, breaking the circuit to the start capacitor, thus disconnecting the start capacitor from the circuit. The armature will hold the contact points open as long as these conditions exist. When the circuit opens, the contact-holding spring pushes the contacts closed again.

Motor Overload Protectors

A motor overload protector is an automatic device that shuts down the motor before the motor windings can be damaged by an overload. The types of protectors used on air conditioning equipment usually protect the compressor motor against excessive current draw, but also function under other abnormal conditions.

As opposed to relays, protectors have no coils; they operate by means of a temperature-responsive metal disc. The motor protector is connected in

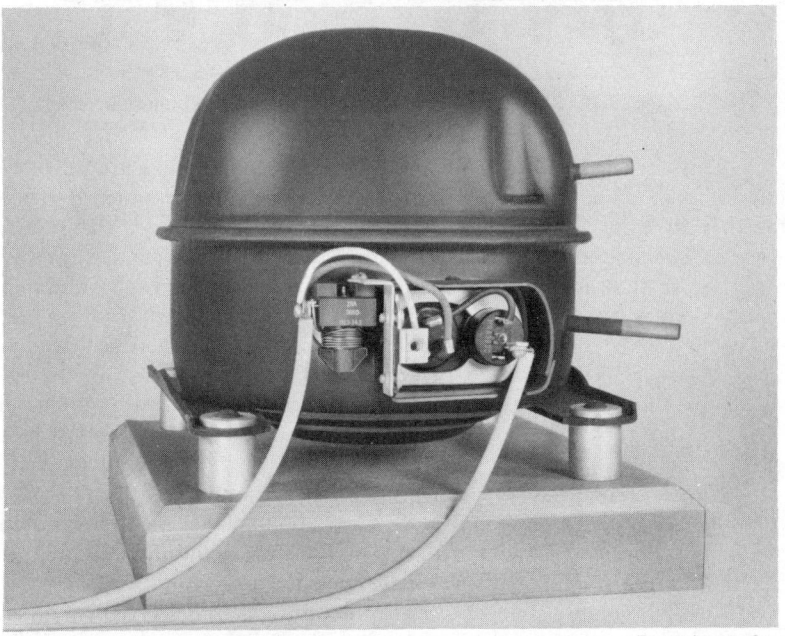

Texas Instru. Co.

Fig. 6-17. Motor overload protector installation.

series with the compressor motor, so that all current going to the compressor motor will also pass through the protector; see Fig. 6-17 for typical installation. Under normal loads, the current going through the protector and the motor is not enough to trip its contacts. During overload, the increased current passing through the protector disc raises the disc temperature to its opening setting. When the disc snaps into its open position, the circuit to the compressor motor is broken, thus shutting off the compressor motor.

Single-Phase Overload Protectors

One type of overload protector used on single-phase motors is comprised of a temperature responsive, snap-acting bimetallic disc, Fig. 6-18. The disc is further controlled by a current sensitive resistance heater connected in series with it. In operation, the disc is normally closed. When a temperature rise causes the disc to snap away from the terminals, the contacts are broken, and the circuit is open.

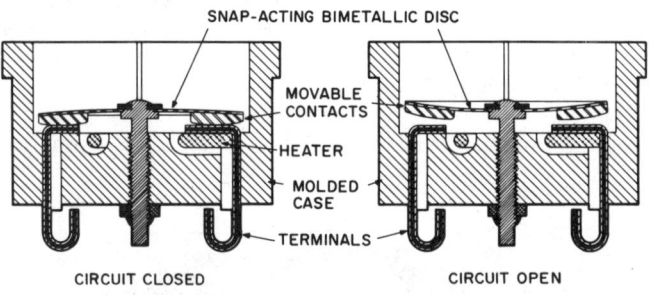

Fig. 6-18. Single-phase overload protector with bimetallic disc.
Texas Instru. Co.

This overload protector operates as follows. The snap-acting BIMETALLIC disc is made up of two strips of dissimilar metals, permanently bonded together to form a single strip. When the disc is heated, these metals expand. The bottom strip has a greater coefficient of expansion than the top piece so that, in case of a temperature rise, it expands more than the top strip does. Since the center of the strip is pinned down, only the strip ends can move. Thus, the greater expansion of the bottom strip will cause the strip to snap into a reverse curvature. This opens the circuit to the motor, thus stopping the motor. The disc can be actuated by a temperature rise due to warm air surrounding the motor, or by excessive motor current.

This type of protector can have either two or three external connection points. When a protector with three points is used, the resistance heater is connected in series with the motor start winding, so that a malfunction in the starting circuit will actuate the protector.

Once the motor shuts off, the temperature eventually returns to normal. The bimetallic disc then snaps closed, and the motor restarts. This

Motors and Controls 1-121

type of protector constantly recycles the motor, as long as the power source remains connected, and until the trouble is remedied.

Another type of overload protector used with single-phase units utilizes only the bimetallic disc. This disc is connected in series with the compressor motor and the power source. The disc senses both motor temperature and current. A malfunction of the motor circuit or excess heat around the motor will open the circuit.

Three-Phase Overload Protectors

Three-phase compressor motors are used in large residential and small commercial applications. The motor overload protector used on three-phase units, Fig. 6-19, comprises a bimetal element and a heater wire. The bimetal element is connected in series with a compressor contactor coil and a low-voltage (24 V) power supply contained in a control panel in the low voltage circuit. The heater wire is connected in series with the compressor contactor and the compressor motor terminals, and is in the high-voltage (line) circuit. The bimetal element senses shell temperature and opens the low-voltage circuit in the event the compressor overheats. The heater wire heats up if there is a malfunction in the compressor circuit, heating the bimetal element and causing it to snap open.

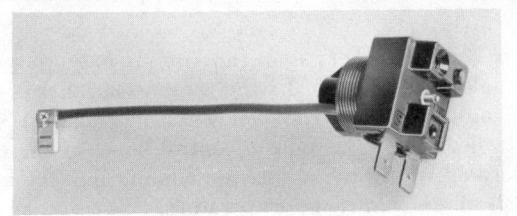

Texas Instru. Co.
Fig. 6-19. Three-phase overload protector.

Temperature Controllers

Temperature controllers, or thermostats, provide automatic control over cooling and/or heating systems. Their use permits presetting the systems for individual comfort. And, once set, the system will maintain the called-for temperature, plus or minus a degree or so, depending on controller tolerance.

Cooling-Only Control

This type of temperature controller acts only to control compressor operation. If the controller is set to a temperature below ambient room temperature, the compressor will operate until the room air temperature reaches the temperature setting of the controller.

1-122   *Basic Air Conditioning*

Ranco
Fig. 6-20. Single-pole, single-throw temperature controller.

One type of controller used for cooling-only purposes utilizes a single-pole, single-throw electrical switch, Fig. 6-20, that is controlled by a bellows and capillary tube. The bellows reacts against spring tension. which is controlled by a knob on the control panel of the unit. The capillary tube is located within the air stream coming into the air conditioning unit, and reacts to ambient room temperature.

When the controller is set for cooling, the room is warm, and the system is OFF, the ambient temperature expands the air in the capillary-bellows arrangement, which causes the bellows to expand. This expansion works against the preset spring tension, turning the single-throw switch ON and furnishing power to the system. Cooling continues until the air flowing past the capillary tube is cool enough to contract the air in the tube. In turn, the bellows contracts, working with spring tension, and turns the single-throw switch OFF, removing system power. The cycle is repetitive as long as the controller is set for cooling and remains live.

Another type of temperature controller used is the combination temperature-controller switch. This unit, shown in Fig. 6-21, incorporates a main cycling switch and an integral auxiliary switch, both controlled through a common dial knob.

Rotating the dial knob from the OFF position immediately turns on the auxiliary fan switch (auxiliary toggle in the illustration). Subsequent dial positions then set the desired temperature settings. The main cycling switch of the controller is a single-pole, single-throw type of switch that operates through a lever by changes of vapor pressures in the capillary bellows arrangement similar to that described before.

Motors and Controls 1-123

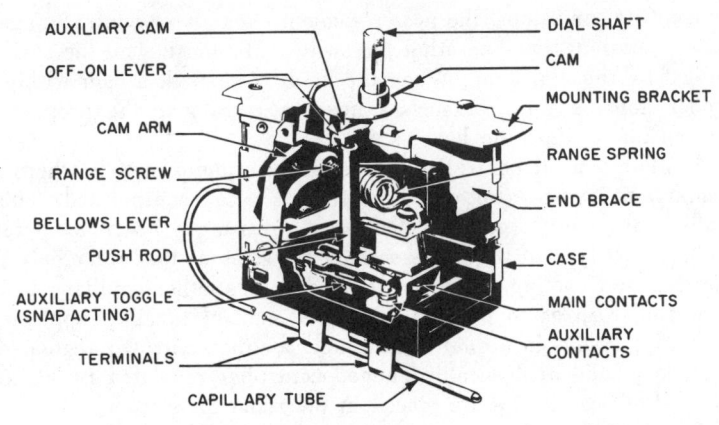

Fig. 6-21. Combination temperature controller switch.
Ranco

Heating and Cooling

Air conditioning units equipped for both heating and cooling (heat pumps) are also governed by a temperature controller that responds to room temperature. In warm weather the controller starts and stops the compressor to maintain the desired level of cooling. In cool weather it automatically reacts to change from the cooling cycle to the heating cycle, and starts the compressor of a typical heat-pump unit to maintain desired heating level.

This type of controller, shown in Fig. 6-22, is basically two single-pole, double-throw switches. These switches operate through a pivoted bellows lever by means of expansion or contraction of a vapor-filled power element (bellows). When the dial knob is set to maintain a specific

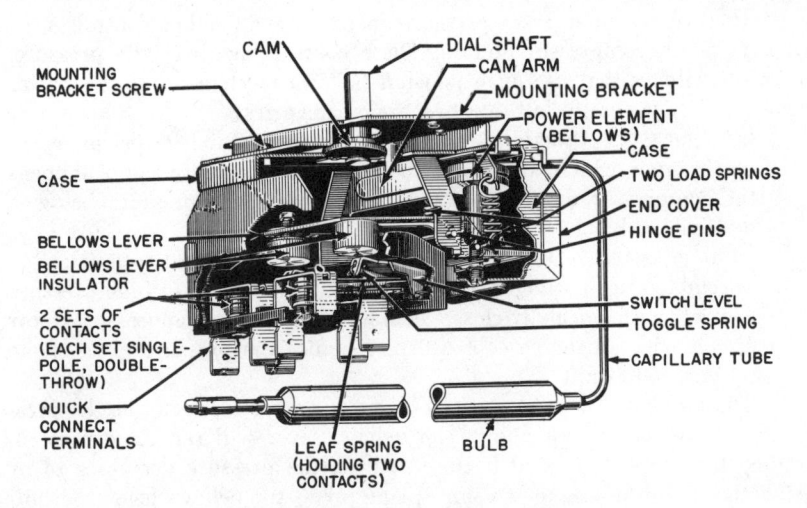

Fig. 6-22. Heating-cooling switch.
Ranco

set of comfort conditions, the power element expands or contracts against the spring tension, thus operating the switch. The tension on the spring is controlled by the dial knob. When the knob is rotated, a cam within the controller housing turns in such a manner as to give the proportional spring tension against the bellows.

Another type of controller is used in heat-pump models where supplementary resistance (hot-wire) heating is also incorporated. This is basically a three-way (cooling, heat-pump, resistance heat) temperature controller. This type of unit is essentially the same as that commonly used for heating and cooling units, but ganged with another capillary bellows type controller so as to provide additional control for the supplemental heating. This results in a dual heating and cooling controller regulated by one capillary, and a physicallly ganged controller regulated by a second capillary. Both capillaries are placed in the room air stream.

Acting upon pre-set temperature conditions, the dual heating and cooling controller, Fig. 6-23, will react to control the individual circuits to either connect the heat pump compressor into the circuit for its cooling operation, or to connect the compressor into the circuit and energize a reversing valve solenoid for heating operation. The third section of the controller will react whenever the room temperature falls further below the pre-set temperature level set for the supplemental heaters.

Electrically this three-stage controller operates as three single-pole, double-throw switches operating off two separate capillary and bellows systems. Their separate operating principles are similar to the operating principles of the heating and cooling temperature controller which was described earlier.

Pressure Controls

There are a number of pressure-operated, snap-action controls used for controlling compressor motors. These controls are basically pressure-actuated switches that use an expansion bellows mechanism as the power element. They generally fall into two basic categories:

LOW PRESSURE (Suction)—for pressures from 7 to 105 psi in selectable ranges which are non-adjustable. This pressure control opens the compressor motor circuit with a decrease of pressure below a desired level.

HIGH PRESSURE (Discharge)—for pressures from 100 to 450 psi in selectable ranges which are non-adjustable. This type of high pressure control, with normally-closed contacts, opens the compressor motor circuit with an increase of pressure before the pressure exceeds the selected safe limit.

The LOW-PRESSURE CONTROL, Fig. 6-24, operates as long as the pressure is above the cut-out point. The bellows lever will remain depressed, holding the snap-action switch closed. When the pressure decreases to or below the cut-out point, the tension spring pivots the bellows lever, permitting the snap-action switch to open.

Motors and Controls 1-125

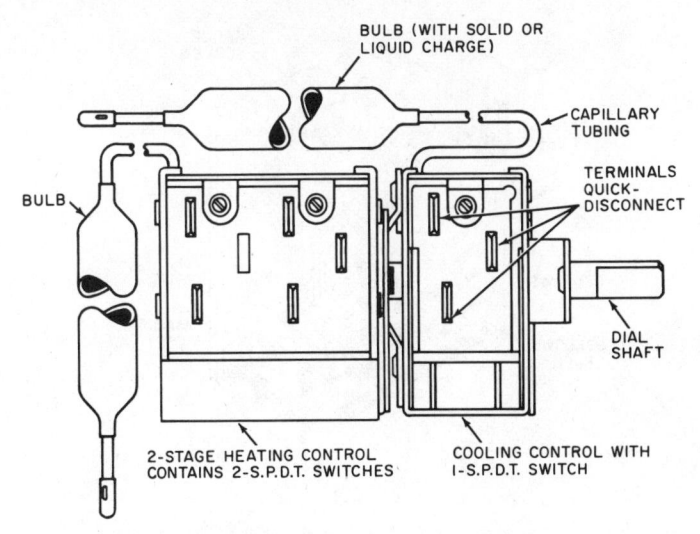

Fig. 6-23. Dual heating-cooling controller.
Ranco

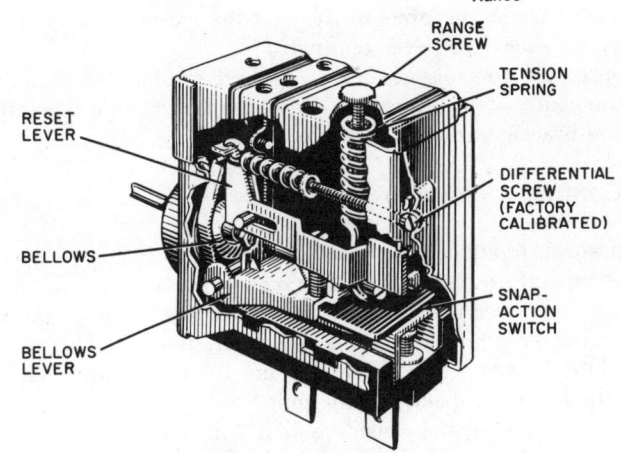

Fig. 6-24. Low-pressure control.
Ranco

In HIGH-PRESSURE CONTROL operation, when the pressure is below the cut-out point, the tension spring, Fig. 6-25, holds the bellows lever in a position that permits the snap-action switch to remain closed. As the pressure increases to or above the cut-out point, the tension spring pivots the bellows lever and the switch opens.

Both the high and low-pressure controls are furnished with either automatic or manual reset. In the MANUAL-RESET TYPE, as the bellows lever moves, the top slides from the action of its own tension spring to block the bellows lever; thus the switch cannot reclose even though the pressure reaches the cut-in point. After the cut-in point is reached, depressing the reset arm allows the switch to close. However, if the reset bar is held in or blocked, the effect then is automatic operation of the compressor.

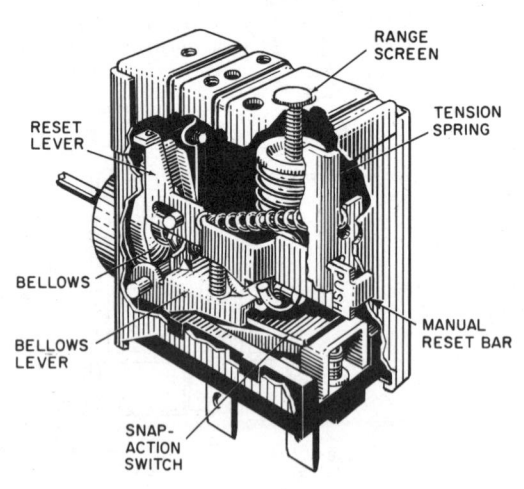

Fig. 6-25. High-pressure control.
Ranco

In pressure controls equipped with AUTOMATIC RESET, the switch will open and close as the pressure reaches the cut-out and cut-in points, respectively, with the reset lever omitted.

Dual-pressure controls are merely paired combinations of low- and high-pressure controls, or two each of either the low or high types, mounted on a common bracket and connected electrically.

De-Icing Controls

When an air-to-air heat pump is in its heating cycle, the outside coil is the low-temperature agent (evaporator) of the system. Whenever the outside coil temperature reaches 32°F, or colder, ice forms on the fins and tubes and accumulates until the coil is completely blocked. This reduces the transfer of heat between the outside coil and the surrounding air, in turn decreasing the heating capacity of the unit.

Once ice forms on the outside coil it will not melt, even if the air temperature around the coil rises, unless the system is stopped for a long time. Since continuous heating is desired, this method is not very practical. The most efficient and quickest method of removing ice from the outside coil is to reverse the refrigerant flow within the system, so that the outside coil once again becomes the condenser. As the outside coil then warms up, the ice is completely removed, and the system's heating capacity is restored.

The de-icer control senses the reduction in heating efficiency and ability of the system, and operates to offset this condition.

Bellows-Operated Control

Electrically, the de-icer control shown in Fig. 6-26 is basically a single-pole, single-throw, snap-acting toggle switch. The switch is controlled

Motors and Controls 1-127

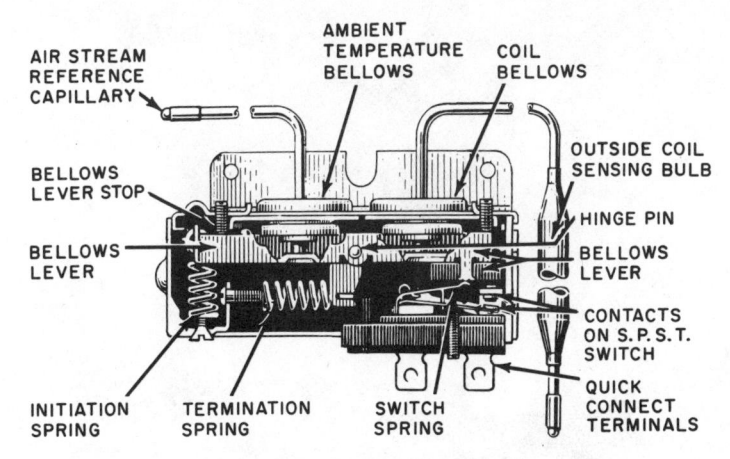

Fig. 6-26. De-icer control.
Ranco

by two bellows systems that oppose each other through a pivoted bellows lever attached at one end to an initiation spring, and connected at the other end to the electrical switch. The lever operates the switch through the difference (spread) in temperature between the two bellows. The latter have unequal effective areas producing different pivot-to-bellows ratios.

One capillary tube, referred to as reference or outside ambient, is in the outside air stream. The other bellows, with a cross-ambient type fill and sensing bulb, is attached to the outside coil. One bellows senses a different temperature than the other bellows and rotates the lever in the direction of the greater force applied by that bellows. This action starts the de-icing cycle, and also terminates it.

The initiation and termination temperatures will vary with the temperature of the outside ambient air. As this air temperature decreases, so do the initiation and termination temperatures.

When the de-ice cycle is initiated, the outside coil (now the condenser) warms up and melts the ice from the coil. The heat pump continues to run in this condition until the coil reaches sufficient temperature (usually 58–60°F on the preset control) to terminate the cycle. The coil temperature stays around 32°F until all ice is removed, then the temperature rises rapidly to the termination temperature. During this cycle the outside fan is stopped to accelerate the de-icing time. Once the termination temperature is reached, the unit reverts to the heating phase.

Air-Operated Control

The air-operated control (air switch) initiates the defrost cycle. When the outside temperature drops to about 40°F or lower, moisture in the air forms as frost and ice on the outside coil, thereby decreasing the efficiency of the unit.

1-128   *Basic Air Conditioning*

Fedders

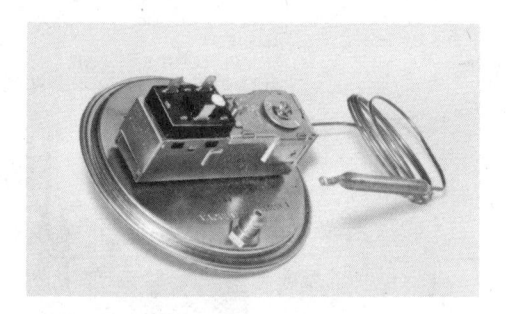

Fig. 6-27. Air switch installation.

    The air switch, Fig. 6-27, senses the restricted air flow caused by the frost and ice build-up, and at a definite factory set air-pressure differential, the control, working in conjunction with the defrost thermostat, initiates the reversing operation of the heat pump to the defrost phase, whereby the outside coil receives hot gas from the compressor. As the temperature in the outside coil increases, the ice and frost melt off the coil. When the outside coil temperature rises to approximately $65°F \pm 7°F$, the defrost thermostat closes, re-starting the fan motor which, in turn, closes the air switch, and the unit goes back into the heating phase if the room thermostat is still calling for heat.

Glossary

Glossary

absolute zero: the zero point on the absolute temperature scale, 459.69 degrees below the zero of the Fahrenheit scale, 273.16 degrees below the zero of the Centigrade scale.

absorption: a process whereby a material extracts one or more substances present in an atmosphere or mixture of gases or liquids; accompanied by physical change, chemical change, or both, of the material.

activated alumina: a form of aluminum oxide which absorbs moisture readily and is used as a drying agent.

adsorption: the action, associated with surface adherence, of a material in extracting one or more substances present in an atmosphere or mixture of gases and liquids, unaccompanied by physical or chemical change.

air, ambient: generally speaking, the air surrounding an object.

air conditioning: the process of treating air so as to control simultaneously its temperature, humidity, cleanliness, and distribution to meet the requirements of the conditioned space.

air cooling: reduction in air temperature due to the subtraction of heat as a result of contact with a medium held at a temperature lower than that of the air. Cooling may be accompanied by moisture addition (evaporation), by moisture extraction (dehumidification), or by no change whatever of moisture content.

air, outdoor: air taken from outdoors and, therefore, not previously circulated through the system.

air, recirculated: return air passed through the conditioner before being again supplied to the conditioned space.

air, return: air returned from conditioned or refrigerated space.

air, saturated: moist air in which the partial pressure of the water vapor is equal to the vapor pressure of water at the existing temperature. This occurs when dry air and saturated water vapor coexist at the same dry-bulb temperature.

air standard: air with a density of 0.075 per cu ft and an absolute viscosity of 0.0379×10^{-5} lb mass per ft-sec. This is substantially equivalent to dry air at 70°F and 29.92 in. Hg barometric pressure.

air, supply: The quantity of air delivered to each or any space in the system, or the total delivered to all spaces in the system.

alternating current (ac): current flow which is constantly changing in amplitude and reversing its direction at regular intervals.

ampere: a unit of intensity of electrical current produced in a conductor by an applied voltage.

ASHRAE: American Society of Heating, Refrigeration, and Air Conditioning Engineers.

barometer: An instrument used for measuring atmospheric pressure.

boiling point: the temperature at which the vapor pressure of a liquid equals the absolute external pressure at the liquid-vapor interface.

British Thermal Unit (Btu): The heat required to raise the temperature of 1 pound of water 1°F.

Btuh: British Thermal Unit per Hour.

buck and boost: stepdown and stepup (transformer).

bypass: a pipe or duct, usually controlled by valve or damper, for conveying a fluid around an element of a system.

calibration: the process of dividing and numbering the scale of an instrument; also of correcting or determining the error on an existing scale, or of evaluating one quantity in terms of the readings of another.

calorie: the heat required to raise the temperature of 1 gram of water 1°C from 4°C to 5°). Mean calorie = 1/100 part of the heat required to raise 1 gram of water from 0° to 100°C.

calorimeter: a device for measuring heat quantities, such as machine capacity, heat of combustion, specific heat, heat leakage, etc. Also a device for measuring quality (or moisture content) of steam or other vapors.

capacity: the usable output of a system or system component in which only losses occurring in the system or component are charged against it.

capacity, heat: the amount of heat necessary to raise the temperature of a given mass one degree. Numerically, the mass multiplied by the specific heat.

capillary tube: in refrigeration, a tube of small internal diameter used as a liquid refrigerant flow control or expansion device between high and low sides; also used to transmit pressure from the sensitive bulb of some temperature controls to the operating element.

Centigrade (C): a thermometric scale in which the freezing point of water is called 0 degrees and its boiling point 100 degrees at normal atmospheric pressure (14.696 psi).

change of state: change from one phase, either solid, liquid, or gas, to another.

charge: amount of refrigerant in a system; to put in the refrigerant charge.

coil: a cooling or heating element made of pipe or tubing.

comfort chart: a chart showing effective temperatures with dry-bulb temperatures and humidities (and sometimes air motion) by which the effects of various air conditions on human comfort may be compared.

compression: in a compression refrigeration system, a process by which the pressure of the refrigerant is increased.

compressor, centrifugal: a nonpositive displacement compressor which depends for pressure rise, at least in part, on centrifugal effect.

compressor, reciprocating: a positive displacement compressor with a piston or pistons moving in a straight line but alternately in opposite directions.

compressor, rotary: one in which compression is attained in a cylinder by rotation of a positive displacement member.

condensation: the process of changing a vapor into liquid by the extraction of heat. Condensation of steam or water vapor is effected in either steam condensers or in dehumidifying coils and the resulting water is called condensate.

condenser: a vessel or arrangement of pipe or tubing in which a vapor is liquefied by removal of heat.

conductor: a material which gives up free electrons easily and offers little opposition to current flow.

control: any device for regulation of a system or component in normal operation, manual or automatic. If automatic, the implication is that it is responsive to changes of pressure, temperature, or some other property whose magnitude is to be regulated.

convection: transfer of heat by movement of a fluid.

critical point: the point at which the liquid and vapor of a substance have identical properties; critical temperature, critical pressure and critical volume are the terms given to the temperature, pressure and volume at the critical point.

current: the flow of elections through a circuit.

current, induced: the electric current produced by moving a conductor in a magnetic field.

cycle: one complete movement of an ac wave from 0 to 360°; on and off operation of an air conditioner (e.g., cycling on and off due to a malfunction).

cycle, refrigeration: the complete course of operation of refrigerant back to a starting point, as evidenced by: a repeated series of thermodynamic processes, or flow through a series of apparatus, or a repeated series of mechanical operations.

dehumidification: the condensation of water vapor from air by cooling below the dewpoint, or the removal of water vapor from air by chemical or physical methods.

density: the ratio of the mass of a specimen of a substance to the volume of the specimen. The mass of a unit volume of a substance. When weight can be used without confusion, as synonymous with mass, density is the weight per unit volume.

desiccant: any absorbent or adsorbent, liquid or solid, that will remove water or water vapor from a material. In a refrigeration circuit the desiccant should be insoluble in the refrigerant.

dewpoint: see temperature, dewpoint.

differential (of a control): the difference between cut-in and cut-out temperatures or pressures.

direct current (dc): current that always flows in the same direction through a circuit.

drier: a device containing a desiccant, placed in the refrigerant circuit; its primary purpose being to collect and hold within the desiccant all water in the system in excess of the amount which can be tolerated in the circulating refrigerant.

duct: a passageway made of sheet metal or other suitable material, not necessarily leak-tight, used for conveying air or other gas at low pressures.

evaporation: change of state from liquid to vapor.

evaporative cooling: the exchange of heat between air and a water spray or wetted surface. The water approaches the wet-bulb temperature of the air, which remains constant.

evaporator: that part of a refrigerating system in which the refrigerant is vaporized to produce refrigeration.

expansion, dry: a process of heat removal by a refrigerant in an evaporator fed by a flow control, responsive to temperature or pressure or both at some point in the evaporator, or to the difference between high and low side pressures, and not to the liquid level in the evaporator.

Fahrenheit (F): a thermometric scale in which 32° denotes freezing point and 212° the boiling point of water under normal pressure at sea level (14.696 psi).

filter: a device to remove solid material from a liquid.

flammability: a material's ability to burn.

flash gas: the gas resulting from the instantaneous evaporation of refrigerant in a pressure-reducing device designed to cool the refrigerant.

flash point: the temperature of combustible material, as oil, at which there is a sufficient vaporization to ignite the vapor, but not sufficient vaporization to support combustion of the material.

fluid: gas or liquid.

freezing point: the temperature at which a given fluid will solidify or freeze upon removal of heat. The freezing point for water is 32°F, or 0°C.

frequency: the number of cycles per second, given in the unit hertz (Hz).

gas: usually a highly superheated vapor which, within acceptable limits of accuracy, satisfies the perfect gas laws.

halide torch: a flame tester generally using alcohol and burning with a blue flame; when the sampling tube draws in halocarbon refrigerant vapor, the color of the flame changes to bright green.

halogen: a non-oxygenated chemical that forms salts by direct union with metals.

heat: the form of energy that is transferred by virtue of a temperature difference.

heat, latent: change of heat content during a change of state. With pure substances, latent heat is absorbed or rejected at constant pressure.

heat, sensible: heat which is associated with a change in temperature, in contrast to a heat interchange in which a change of state (latent heat) occurs.

heat, specific: the ratio of the quantity of heat required to raise the temperature of a given mass of any substance one deg to the quantity required to raise the temperature of an equal mass of a standard substance (usually water at 59°F) one deg.

heat exchanger: a device specifically designed to transfer heat between two physically separated fluids.

heat of fusion: latent heat involved in the change between solid and liquid states.

heat of vaporization: latent heat involved in the change between liquid and vapor states.

high side: parts of a refrigerating system subjected to condenser pressure or higher.

horsepower (hp): unit of energy; equivalent to 746 watts, 2,545 Btuh.

humidifier: a device to add moisture to air.

humidify: to add water vapor to the atmosphere; to add water vapor or moisture to any material.

humidity: water vapor within a given space.

humidity, relative: approximately, the ratio of the partial pressure or density of the water vapor in the air, to the saturation pressure or density of water vapor at the same temperature.

humidity, specific: weight of water vapor (steam) associated with one pound of dry air; also called the humidity ratio.

insulation, thermal: a material having a relatively high resistance to heat flow, and used principally to retard the flow of heat.

insulator: a material which does not give up free electrons easily and offers great opposition to current flow.

liquid line: the tube or pipe carrying the refrigerant liquid from the condenser or receiver of a refrigerating system to a pressure-reducing device.

load, estimated design: in a heating or cooling system, the sum of the useful heat transfer, plus heat transfer from or to the connected piping, plus heat transfer occuring in any auxiliary apparatus connected to the system.

low side: parts of a refrigerating system at or below evaporator pressure.

mass: the quantity of matter in a body as measured by the ratio of the force required to produce given acceleration to the acceleration.

Ohm: basic unit of resistance measure equal to that resistance which allows 1 ampere of current flow when an emf (potential) of 1 volt is applied across the resistance.

oil separator: a device for separating oil and oil vapor from the refrigerant, usually installed in the compressor discharge line.

power: the rate of performing work. Common units are horsepower, Btuh, and watts.

power factor: the figure which indicates what portion of the current delivered to the motor is used to do work.

preheating: in air conditioning, to heat the air in advance of other processes.

pressure: the normal force exerted by a homogeneous liquid or gas, per unit of area, on the wall of its container.

pressure, absolute: pressure referred to that of a perfect vacuum. It is the sum of gauge pressure and atmospheric pressure.

pressure, atmospheric: the pressure due to the weight of the atmosphere. It is the pressure indicated by a barometer. Standard atmospheric pressure or standard atmosphere is the pressure of 76 cm of mercury having a density of 13.5951 grams per cu cm, under standard gravity of 980.665 cm per sec^2. It is equivalent to 14.696 psi or 29.921 in. of mercury at 32°F.

pressure, static: the normal force per unit area that would be exerted by a moving fluid on a small body immersed in it if the body were carried along with the fluid. Practically, it is the normal force per unit area at a small hole in a wall of the duct through which the fluid flows or on the surface of a stationary tube at a point where the disturbances, created by inserting the tube, cancel out.

pressure, vapor: the pressure exerted by a vapor. If a vapor is kept in confinement over its liquid so that the vapor can accumulate above the liquid, the temperature being held constant, the vapor pressure approaches a fixed limit called the maximum, or saturated, vapor pressure, dependent only on the temperature and the liquid. The term vapor pressure is sometimes used as synonymous with saturated vapor pressure.

psychrometry: the branch of physics relating to the measurement or determination of atmospheric conditions, particularly regarding moisture mixed with the air.

purging: the act of blowing out gas from a refrigerant-containing vessel, usually for the purpose of removing noncondensables.

receiver: a vessel permanently connected to a system by inlet and outlet pipes for the storage of condensed refrigerant.

refrigerant: a substance which produces a refrigerating effect by its absorption of heat while expanding or vaporizing.

refrigerating system, mechanical: a refrigerating system employing a mechanical compression device to remove the low-pressure refrigerant enclosed in the low-pressure side and deliver it to the high-pressure side of the system.

resistance: opposition offered by a material to the flow of current.

rotor: the rotating section of a motor.

saturation: the condition for coexistence in stable equilibrium of a vapor and liquid or a vapor and solid phase of the same substance. Example: steam over the water from which it is being generated.

serpentining: doubling the tube in an evaporator back upon itself several times to increase length in a short space.

sight glass: glass tube used to indicate the liquid level in pipes, tanks, bearings and similar equipment.

silica gel: a form of silicon dioxide which adsorbs moisture readily and is used as a drying agent.

sine wave: a curve of theoretically pure ac.

solenoid: an electromagnet with an energizing coil around a plunger. The plunger moves when current is applied to the coil.

stator: the stationary section (field coil) of a motor.

sub-base: separate plate in a thermostat that holds all the components.

subcooling: the process of cooling refrigerant below condensing temperature, for a given pressure; also, cooling a liquid below its freezing point, where it can exist only in a state of unstable equilibrium.

temperature: the thermal state of matter with reference to its tendency to communicate heat to matter in contact with it. If no heat flows upon contact, there is no difference in temperature.

temperature, critical: the saturation temperature corresponding to the critical state of the substance at which the properties of the liquid and vapor are identical.

temperature, dewpoint: the temperature at which the condensation of water vapor in a space begins for a given state of humidity and pressure as the temperature of the vapor is reduced. The temperature corresponding to saturation (100 percent relative humidity) for a given absolute humidity at constant pressure.

temperature, dry-bulb: the temperature of a gas or mixture of gases indicated by an accurate thermometer after correction for radiation.

temperature, effective: an arbitrary index which combines into a single value the effect of temperature, humidity, and air movement on the sensation of warmth or cold felt by the human body. The numerical value is that of the temperature of still, saturated air which would induce an identical sensation.

temperature, wet-bulb: thermodynamic wet-bulb temperature is the temperature at which liquid or solid water, by evaporating into the air, can bring the air to saturation adiabatically at the same temperature. Wet-bulb temperature (without qualification) is the temperature indicated by a wet-bulb psychrometer constructed and used according to specifications.

thermostat: automatic heating–cooling control.

ton of refrigeration: a useful refrigerating effect equal to 12,000 Btu per hour; 200 Btu per min.

vapor, superheated: vapor at a temperature which is higher than the saturation temperature (i.e. boiling point) at the existing pressure.

triple point: the temperature at which three phases of given substance (solid, liquid, and gas) exist in equilibrium at atmospheric pressure.

vapor: a gas, particularly one near to equilibrium with the liquid phase of the substance, which does not follow the gas laws. Usually used for a refrigerant, and in general for any gas below the critical temperature.

voltage: electromotive force (emf). It is a force which, if applied to a closed circuit, will produce a current in the circuit.

zone, comfort: average—the range of effective temperatures over which the majority (50% or more) of adults feel comfortable. Extreme—the range over which one or more feel comfortable.

Index

Index

Absolute pressure, 1-12
Absolute zero, 1-5 to 1-6
Accumulators, 1-70
Air conditioning, 1-3
    principles, 1-29 to 1-31
Air switches, 1-128
Amperage, 1-17 to 1-18
    nameplate full-load, 1-20 to 1-21
Aneroid barometers, 1-97
Anti-slug devices, 1-54 to 1-55
Atmospheric pressure, 1-12

Barometers
    aneroid, 1-97
    manometers, 1-97
    mercury, 1-95 to 1-97
Bellows pressure gauges, 1-98 to 1-99
Boiling point
    and pressure, 1-14 to 1-15
    in refrigerating cycle, 1-32
Bourdon tube pressure gauges, 1-99 to 1-100
Boyle's Law, 1-13 to 1-14

Capacitive devices, 1-26
Capacitors, 1-26
    run, 1-114 to 1-115
    start, 1-113 to 1-114
Capacity control
    cylinder bypass, 1-54 to 1-56
    cylinder unloading, 1-54 to 1-55
    hot-gas-bypass, 1-56 to 1-57
Capillary tubes, 1-34
    operation, 1-69 to 1-70
Centigrade (Celsius) temperature scale, 1-5 to 1-6
Change of state, 1-8 to 1-9
Charges
    all-purpose, 1-76
    gas, 1-74 to 1-75
    liquid, 1-75
    liquid cross, 1-75

Charles' Law, 1-14
Check valves, 1-92 to 1-93
Circuits, electrical, 1-27 to 1-28
    phase, 1-22 to 1-25
Compound pressure gauges, 1-100
Compression stages, 1-47
Compressor mounting, 1-47 to 1-48
Compressor seals, 1-49 to 1-50
Compressor system, 1-45
Compressor valves, 1-49 to 1-50
Compressors, 1-33 to 1-34
    backward-curved impeller, 1-60
    capacity control, 1-54 to 1-57
    centrifugal, 1-59 to 1-61
    dynamic, 1-45
    lubrication, 1-52
    positive displacement, 1-45
    radial blade, 1-59 to 1-60
    reciprocating, 1-46 to 1-47
    rotary, 1-57 to 1-58
    rotating blade, 1-58 to 1-59
    slugging, 1-52 to 1-54
Condensation
    and pressure, 1-16
    in refrigerating cycle, 1-32
Condensers, 1-33 to 1-35, 1-61 to 1-62, 1-68
    air-cooled, 1-62 to 1-65, 1-84 to 1-85
    in refrigerating cycle, 1-33
    water-cooled, 1-62 to 1-63, 1-65 to 1-68
Conduction, 1-10
Conductors, 1-26 to 1-27
Constant-pressure expansion valve, 1-70
Convection, 1-11
Current, 1-17 to 1-19
    waveforms, 1-19
Cycles, 1-21 to 1-22

De-icing controls
    air operated, 1-127 to 1-128
    bellows operated, 1-126 to 1-127

1-141

Index

Desiccants, 1-101 to 1-104
Diaphragm operated pressure
    gauges, 1-97 to 1-98
Distributors, 1-78 to 1-79
    centrifugal, 1-80 to 1-81
    pressure drop, 1-79 to 1-80
    Venturi-flo, 1-79 to 1-80

Efficiency, 1-25 to 1-26
Electric relays
    current, 1-117 to 1-118
    potential, 1-118 to 1-119
    solenoid, 1-115
    starting, 1-116
    thermal, 1-115 to 1-116
Electrical theory, 1-16 to 1-19
Equalizers
    external, 1-78 to 1-79
    internal, 1-76 to 1-77
Evaporation
    in cooling, 1-31
    and pressures, 1-14 to 1-15
Evaporators, 1-38 to 1-41
    coil arrangements, 1-41
    dry, 1-42 to 1-44
    fin arrangements, 1-41
    flooded, 1-41 to 1-42
    performance chart, 1-44
    pressure regulators, 1-83 to 1-84
    in refrigerating cycle, 1-33
Expansion valves, 1-70 to 1-71

Fahrenheit temperature scale, 1-5 to
    1-6
Filter-driers, 1-101 to 1-104
    contaminants, 1-102
    replaceable cores, 1-104
Float valves, 1-81 to 1-82
Frequency
    nameplate, 1-21 to 1-22
    relation to period, 1-22
Fusion, heat of, 1-10

Gauge pressure, 1-12 to 1-13
Gauges, *see* Pressure gauges

Heat, 1-6
    latent, 1-8 to 1-10
    measurement, 1-7
    specific, 1-8 to 1-9
Heat of fusion, **1-10**
Heat of vaporization, 1-10

Heat transfer, 1-10 to 1-11
Hertz, 1-21
Humidity, 1-29 to 1-31

Inductive devices, 1-26 to 1-27
Insulators, 1-26 to 1-27

Kelvin temperature scale, 1-5 to 1-6
Kinetic energy, 1-6

Latent heat, 1-8 to 1-9
Latent heat of fusion, 1-10
Latent heat of vaporization, 1-10
Liquid slugging, 1-52 to 1-54
Lubrication, compressors, 1-52

McLeod gauge, 1-100 to 1-102
Manometers, U-shaped, 1-96 to 1-97
Matter, 1-3 to 1-4, 1-16
Mercury barometers, 1-95 to 1-97
Motors
    automatic controls, 1-112 to 1-119
    fan, 1-105 to 1-106
    overload protectors, 1-119 to 1-121
    polyphase, 1-111 to 1-112
    single-phase, 1-108 to 1-111

Nameplates, 1-19 to 1-22

Ohm's Law, 1-18 to 1-19
Overload protectors, 1-119 to 1-120
    single-phase, 1-120 to 1-121
    three-phase, 1-121

Parallel circuits, 1-28
Phase, 1-22 to 1-23
    two-phase systems, 1-23 to 1-24
    three-phase systems, 1-24 to 1-25
Power factor, 1-25
Pressure, 1-11
    absolute, 1-12
    atmospheric, 1-12
    effects of, 1-13 to 1-16
    gauge, 1-12 to 1-13
Pressure controls
    automatic reset, 1-126
    high pressure, 1-124 to 1-126
    low pressure, 1-124 to 1-125
    manual reset, 1-125

Index

Pressure gauges, 1-95 to 1-101
  barometers, 1-95 to 1-97
  bellows, 1-98 to 1-99
  Bourdon-tube, 1-99 to 1-100
  compound, 1-100
  diaphragm, 1-97 to 1-98
  McLeod, 1-100 to 1-101
Pressure regulators
  air-cooled condenser, 1-84 to 1-85
  evaporator, 1-83 to 1-84
  suction, 1-83 to 1-84
  water-cooled condenser, 1-85 to 1-86

Radiation, 1-11
Rankine temperature scale, 1-5 to 1-6
Receivers, liquid, 1-68 to 1-69
Refrigerant metering devices, 1-69 to 1-82
Refrigerants, 1-37 to 1-40, 1-74
Refrigerating cycle, 1-32 to 1-37
Refrigeration, 1-3
Relief valves, 1-94 to 1-95
Resistive devices, 1-26
Reversing valves, 1-89 to 1-91
Rotors, 1-105 to 1-106
  squirrel-cage, 1-107 to 1-108

Seals
  diaphragm, 1-50, 1-52
  packing gland, 1-49 to 1-51
  rotary, 1-50
  stationary bellows, 1-50 to 1-51
Semiconductors, 1-27
Sensible heat, 1-8 to 1-9
Series circuits, 1-27 to 1-28
Serpentining, 1-40 to 1-41, 1-43
Shut-off valves
  manual, 1-86 to 1-87
  solenoid, 1-87 to 1-89
Slugging, *see* Liquid slugging
Specific heat, 1-8 to 1-9

Stators, 1-105 to 1-106
  two-pole, 1-106 to 1-107
Switches, air, 1-128
Synchronous speed, 1-107 to 1-108

Temperature, 1-4 to 1-5
Temperature scales, 1-5 to 1-6
Thermal bulbs, 1-71 to 1-73
Thermometers, 1-5 to 1-6
Thermostatic expansion valves, 1-70 to 1-71
  all-purpose charge, 1-76
  gas charge, 1-74 to 1-75
  liquid charge, 1-75
  liquid cross charge, 1-75
  pressures, 1-72 to 1-73
Thermostats
  combination switches, 1-123 to 1-124
  cooling only, 1-121 to 1-123
  heating/cooling, 1-123 to 1-125
  single-pole, 1-122

Vacuum, 1-11
Valves, 1-49 to 1-50
  check, 1-92 to 1-93
  expansion, 1-70 to 1-76
  with external equalizer, 1-78
  float, 1-81 to 1-82
  with internal equalizer, 1-76 to 1-77
  relief, 1-94 to 1-95
  reversing, 1-89 to 1-91
  shut-off, 1-86 to 1-89
Vaporization
  heat of, 1-10
  and pressure, 1-14 to 1-15
  in refrigerating cycle, 1-32 to 1-35
Ventilation, 1-29
Voltage, 1-16 to 1-17
  input, 1-20 to 1-21
  nameplate, 1-19 to 1-20